Climate of Hope: New Strategies for Stabilizing the World's Atmosphere

CHRISTOPHER FLAVIN
ODIL TUNALI

CHAMPLAIN COLLEGE

Jane A. Peterson, *Editor*

WORLDWATCH PAPER 130
June 1996

THE WORLDWATCH INSTITUTE is an independent, nonprofit environmental research organization in Washington, D.C. Its mission is to foster a sustainable society in which human needs are met in ways that do not threaten the health of the natural environment or future generations. To this end, the Institute conducts interdisciplinary research on emerging global issues, the results of which are published and disseminated to decision makers and the media.

FINANCIAL SUPPORT is provided by Carolyn Foundation, the Nathan Cummings Foundation, the Geraldine R. Dodge Foundation, the Energy Foundation, The Ford Foundation, the George Gund Foundation, The William and Flora Hewlett Foundation, W. Alton Jones Foundation, John D. and Catherine T. MacArthur Foundation, Andrew W. Mellon Foundation, The Curtis and Edith Munson Foundation, Edward John Noble Foundation, The Pew Charitable Trusts, Lynn R. and Karl E. Prickett Fund, Rockefeller Brothers Fund, Rockefeller Financial Services, Surdna Foundation, Turner Foundation, U.N. Population Fund, Wallace Genetic Foundation, Wallace Global Fund, Weeden Foundation, and the Winslow Foundation.

PUBLICATIONS of the Institute include the annual *State of the World*, which is now published in 27 languages; *Vital Signs*, an annual compendium of global trends that are shaping our future; the *Environmental Alert* book series; *World Watch* magazine; and the Worldwatch Papers. For more information on Worldwatch publications, write: Worldwatch Institute, 1776 Massachusetts Ave., NW, Washington, DC 20036; or fax 202-296-7365; or see back pages.

THE WORLDWATCH PAPERS provide in-depth, quantitative and qualitative analysis of the major issues affecting prospects for a sustainable society. The Papers are written by members of the Worldwatch Institute research staff and reviewed by experts in the field. Published in five languages, they have been used as concise and authoritative references by governments, nongovernmental organizations, and educational institutions worldwide. For a partial list of available Papers, see back pages.

DATA from all graphs and tables contained in this Paper are available on 3 1/2" Macintosh or IBM-compatible computer disks. The disks also include data from the *State of the World* series, *Vital Signs*, *Environmental Alert* book series, Worldwatch Papers, and *World Watch* magazine. Each yearly subscription includes a mid-year update, and *Vital Signs* and *State of the World* as they are published. The disk is formatted for Lotus 1-2-3, and can be used with Quattro Pro, Excel, SuperCalc, and many other spreadsheets. To order, see back pages.

© Worldwatch Institute, 1996
Library of Congress Catalog Number 96-60648
ISBN 1-878071-32-7

Printed on 100-percent non-chlorine bleached, partially recycled paper.

Table of Contents

Introduction 5

The Evidence Mounts 10

Climate Shocks 20

Reducing Greenhouse Gas Emissions 31

Stabilizing the Climate 42

Rio, Berlin, and Beyond 53

Notes 70

Tables and Figures

Table 1: *Weather-related Disasters with Damages* 27

Table 2: *Carbon Emissions from Fossil Fuel Burning* 34

Table 3: *International Climate Alliances, Berlin, 1995* 64

Figure 1: *Global Average Surface Temperature, 1866–1995* 13

Figure 2: *Temperature Projections with and without Sulfate* 14

Figure 3: *Local Temperature and Atmospheric Carbon Dioxide* 19

Figure 4: *Economic Losses from Weather-related Disasters* 30

Figure 5: *World Carbon Emissions from Fossil Fuel Burning* 33

Figure 6: *Atmospheric Carbon Dioxide Concentrations* 44

Figure 7: *World Carbon Emissions, 1900–95* 45

For one-time academic use of this material, please contact Customer Service, Copyright Clearance Center, at (508) 750-8400 (phone), or (508) 750-4744 (fax), or write to CCC, 222 Rosewood Drive, Danvers, MA 01923. Nonacademic users, please call the Worldwatch Institute's Communication Department at (202) 452-1992, x520, or fax your request to (202) 296-7365.

The views expressed are those of the author(s) and do not necessarily represent those of the Worldwatch Institute, its directors, officers, or staff, or of its funding organizations.

ACKNOWLEDGMENTS: We are grateful to Michael Grubb, Bill Moomaw, Michael Oppenheimer, Christoph Bals, and Kilaparti Ramakrishna, as well as our colleagues Lester Brown, Hilary French, Bill Mansfield, David Malin Roodman, and Jim Perry for their valuable comments on preliminary drafts of this paper. David also ably assisted with the translation of complicated climate data into easy-to-read figures. In addition, we thank our Worldwatch colleagues: Chris Bright, Suzanne Clift, Jennifer Seher, Denise Byers Thomma, Lori Ann Baldwin, Laura Malinowski, and Anjali Acharya for their invaluable help with research, editing, and production, and most of all, for their patience. Finally, we would like to thank our editor, Jane Peterson, for her insight and criticial reviews.

CHRISTOPHER FLAVIN is Vice-President for Research at the Worldwatch Institute. His research and writing focus on international energy and climate policy. He is co-author of *Power Surge: Guide to the Coming Energy Revolution* (W.W. Norton & Co.: 1994). He has participated in many United Nations conferences dealing with climate issues, including the Earth Summit in Rio de Janeiro in 1992 and the first Conference of the Parties to the Framework Convention on Climate Change in Berlin in 1995. He serves on the boards of the Business Council for a Sustainable Energy Future and the American Wind Energy Association.

ODIL TUNALI is a Staff Researcher at Worldwatch Institute, where she studies global climate change, energy, and transportation issues. She is co-author of the Institute's *Vital Signs 1995* and *Vital Signs 1996* reports, and contributes regularly to *World Watch* magazine. She represented the Institute at the Second United Nations Conference on Human Settlements (Habitat II) in Istanbul in June 1996.

Introduction

During the four years since the Framework Convention on Climate Change was signed in Rio de Janeiro in 1992, public discussion of the issue has been dominated by a vitriolic and increasingly sterile debate over whether the earth is indeed warming. In 1996, however, climate deliberations are entering a new phase—one that is marked by new scientific evidence of the climatic risks the world faces, and rising optimism that the means are at hand to begin the historic shift away from fossil fuels.

Most scientists now believe that human-induced climate change has already begun. This conclusion is supported not only by a steady rise in global average temperatures but by evidence of local changes as well—including shifts in the timing of the seasons, the melting of many glaciers, and the rapid migration of some plant and animal species. The U.N. panel of scientists that advises governments on climate change concluded in 1995 that increasing levels of greenhouse gases were almost certainly affecting recent climate patterns. According to climatologist Tom Wigley of the U.S. National Center for Atmospheric Research, recent advances in climate science "mark a turning point...in our ability to understand past changes and predict the future."[1]

Both natural evolution and human social evolution have been shaped by climate change. At times, climatic shifts have wiped out huge areas of forests and other ecosystems, eliminating hundreds of species. And some ancient civilizations disappeared after changing weather undermined their water supplies and food production. Although climate change is not new, such rapid change is. With atmospheric concentrations of carbon dioxide now at their highest level in 150,000 years—and still increasing—the world will likely face a rate of change in the next several decades that exceeds "natural" rates by a factor of ten.

Scientists believe that the coming period of rapid climate change is likely to be erratic, disruptive, and unpredictable. Local weather patterns may shift suddenly and dangerously. According to recent studies, the incidence of floods, droughts, fires, and heat outbreaks will probably increase as global temperatures rise. In fact, some of these changes may already have begun.[2]

At a time when the world is adding 87 million people—more than the entire population of Germany—each year, and already suffering from unprecedented loss of forests and other natural systems, we are not well positioned to cope with such disruptions. Chronic water shortages now plague 80 countries, home to 40 percent of the world's population, and demand for food is pushing against production capacity in many regions. Weather extremes have reduced the harvests of China and the United States twice already in the 1990s. Hotter, drier weather conditions could undermine a precarious global food balance. And recent studies indicate that much of the damage from climate change is likely to fall on developing countries that are least able to cope.

Besides agriculture, other industries that are especially vulnerable to climate change include fishing, forest products, and insurance. The global insurance industry paid out $57 billion in weather-related claims between 1990 and 1995, compared to $17 billion for the entire preceding decade, leading some insurance company executives to worry that the weather extremes likely to accompany climate change could be devastating. Franklin Nutter, president of the Reinsurance Association of America, sums up the threat: "The insurance industry is first in line to be affected by climate change...it could bankrupt the industry."[3]

The objective of the Rio climate convention is clearly stated: to stabilize the world's atmosphere so as to prevent dangerous interference with global weather patterns. So far, however, little progress has been made toward that ambitious goal. Diplomats from more than 150 nations have labored for four years over an array of mind-numbing technical and legal issues, but they have only begun to spur the concrete changes in governmental and business priorities that ultimately are needed to stabilize the climate.

Climate change is likely to be erratic, disruptive, and unpredictable.

Combustion of fossil fuels has continued to grow, releasing ever-greater amounts of carbon dioxide (CO_2), the leading greenhouse gas. Global carbon emissions from fossil fuels reached a record of just over 6 billion tons in 1995. Roughly 1.6 billion tons of additional carbon are emitted from forest clearing each year. Meanwhile, of the 35 industrial countries that committed themselves under the treaty to hold their greenhouse gas emissions to 1990 levels in the year 2000, only about half now appear likely to meet that target. In some developing countries, emissions are on course to nearly double between 1990 and 2000.

The world's failure to heed the warnings of scientists appears to have several roots. The first is the complexity of the forces driving climate change. Since carbon dioxide is a virtually inevitable byproduct of a fossil-fuel-based energy system, efforts to stabilize the climate will at some point require a fundamental revamping of that system. Exactly how to do this and what it will cost have been subjects of considerable uncertainty and vehement debate.

The second problem is the geographic breadth and multigenerational time frame of global climate change—challenging societies to work cooperatively to protect distant populations as well as those yet unborn. The third problem is uncertainty over the mix of policies that constitute the most effective and economical means to encourage less use of fossil fuels—a subject of considerable disagreement among policy analysts.

The recent lack of progress on climate policy is less

about complex analytical questions, however, than hardball politics. Among the organized groups now actively engaged in shaping climate policy are fossil fuel producers such as the Exxon Corporation and the Government of Kuwait, automobile and petrochemical companies that use fossil fuels, and a loose group of fossil-fuel-dependent countries, such as the United States and Australia. Many of these nations and industries still seem to hope that their multi-million-dollar public relations efforts will make the whole issue go away. Even in Germany, which has led recent international efforts to slow climate change, the auto and chemical industries managed to block an effort by the government in early 1996 to introduce new energy taxes.

Still, despite the frustrating slowness of the climate policy process so far, the world may soon be ready to make greater strides in slowing climate change. The rapid pace of recent advances in electronics, synthetic materials, and biotechnology offers hope. These technologies may allow the speedy development of a highly efficient energy economy that substitutes ingenuity for the brute force that characterizes today's more primitive system. Most of today's central power plants turn just one-third of the fuel they consume into electricity—and they are efficient compared to most lightbulbs and automobiles, which effectively employ less than 15 percent of the energy supplied. New technologies can drastically increase these efficiencies and also allow energy needs to be met by harnessing the ambient energy of sunlight and wind.

Recent studies show the potential for a cost-effective transition to a new kind of energy system. Just as the vast effort to replace horsepower with steam engines at the end of the last century drove economic growth, so can the shift to a highly efficient, low-carbon energy system revitalize the global economy in the early part of the next century.

Although early debates about climate policy were dominated by the industries that contribute most to the problem, new interest groups have now joined the fray, representing both those most vulnerable to the effects of climate change and those with an economic interest in the new technologies that could replace fossil fuels. The Alliance of Small Island States (AOSIS)—composed of 36 nations that are threatened by rising seas—now lobbies actively for stronger climate policies. They have been joined by several insurance companies that are exposed to weather-related damages, as well as firms in the solar energy, wind power, and energy efficiency businesses.

The support of these governments and industries for strong climate policies has fostered a more hopeful atmosphere at recent negotiations. They point out that major investments in a less carbon-intensive energy system will only be made if policies are changed, particularly in energy and transportation. One role for the climate convention is to galvanize policymakers at every level—from international organizations such as the World Bank to national, regional, and even local administrations, which play an important role in transportation planning and the regulation of energy utilities.

Some analysts and policymakers have pressed for a universal system of carbon dioxide taxes to restrain emissions; others espouse a global emissions trading system. Although no such "global" policy is likely to be implemented anytime soon, it will almost certainly be needed at some point. Meanwhile, the convention can be used to accelerate narrower reforms such as tax incentives for energy efficiency, fuel economy standards for automobiles, the removal of subsidies for forest clearing and driving, and the transfer of new technologies to developing countries.

At the first Conference of the Parties to the climate convention held in Berlin in 1995, governments adopted the "Berlin Mandate," which requires that an amendment or protocol to the treaty be adopted by 1997. This new commitment is to include additional targets and timetables extending beyond the year 2000, as well as specific policies and measures to be adopted by treaty members. The strength of these new agreements will have a profound influence on the world's ability to prevent catastrophic rates of climate change in the next few decades.

The time for action is at hand. With the global economy booming, particularly in developing nations, industries and governments are currently installing large numbers of factories, power plants, roads, and buildings that could contribute to greenhouse gas emissions for many decades to come. Unless the world soon shifts to a new and more sustainable path of energy development, later efforts to stabilize the climate will be far more difficult and expensive—and too late to prevent some of the most wrenching damage. Industrial countries, which unwittingly created the climate problem in the first place and control the technologies that ultimately can solve it, have a clear responsibility to lead the way forward.

The Evidence Mounts

As recently as 1992, the U.N.'s Intergovernmental Panel on Climate Change (IPCC), a group of 2,500 scientists that advises international climate negotiators, predicted that it would be at least a decade before they could confirm a definitive link between current weather trends and the human-induced buildup of greenhouse gases. Yet in late 1995, after a detailed review of several new scientific studies, the IPCC concluded that "a pattern of climatic response to human activities is identifiable in the climatological record."[4]

As this unexpectedly early finding suggests, scientific understanding of the world's climate is advancing exponentially. Although scientists have worked for decades to better understand daily weather patterns, until the late 1980s, the science of long-term global climate trends was a backwater, attracting just a few hundred scientists worldwide. Since then, the field has exploded as governments have recognized the dangers of global climate change and devoted hundreds of millions of dollars to enhancing understanding of it. Among the results: more detailed records of weather trends, relying on surface, atmospheric, and satellite-based instruments; analysis of data from ice cores and geological deposits that enable better understanding of climate trends in the more distant past; and ever more sophisticated computer models that use thousands of equations to mimic the atmosphere and oceans.

Exactly a century ago, in 1896, the Swedish chemist Svante Arrhenius calculated that carbon dioxide released from the burning of coal would significantly raise atmospheric temperatures over time. He understood that carbon dioxide is a greenhouse gas, letting in visible light and other radiation from the sun, yet, like a blanket, trapping heat near the earth's surface. At first, the idea that a few billion human beings could alter the composition of the atmosphere itself must have seemed absurd. But this mass of air turns out to be surprisingly thin and delicate—most of it extending just 50 kilometers from the surface. If the earth were the size of an apple, the atmosphere would be roughly as thick as its skin.[5]

A century after Arrhenius' paper was published, the concentration of carbon dioxide in the atmosphere reached 361 parts per million (ppm)—higher than at any time in the past 150,000 years, and 30 percent above the level that prevailed before fossil fuel burning began with the industrial revolution. By burning fossil fuels and clearing forests, we are converting oxygen to carbon dioxide, in effect reversing the process that living organisms first used to create our oxygen-containing atmosphere billions of years ago. And the process is accelerating. Roughly two-thirds of the buildup of carbon dioxide has occurred since combustion of fossil fuels began to skyrocket after World War II.[6]

In the 1980s, scientists first detected increases in other greenhouse gases that are less abundant but more powerful than carbon dioxide—notably methane (CH_4), nitrous oxide (N_20), chlorofluorocarbons (CFCs), and their substitutes, hydrochlorofluorocarbons (HCFCs) and hydrofluorocarbons (HFCs). Concentrations of methane, a natural constituent of the atmosphere, have increased by 145 percent since pre-industrial times, mainly as a result of emissions from rice paddies, livestock, landfills, and fossil fuel extraction and processing. The third largest contributor to climate warming—CFCs and HCFCs—are manmade chemicals used as coolants and for industrial processes. They are already being phased out due to their contribution to ozone depletion, but because they last a long time in the atmosphere, their concentration will continue to grow for at least the next decade. These other gases increase the heat-trapping effect of CO_2 alone by 50 percent. Altogether, the added greenhouse gases trap 2.5 watts per square meter of the earth's surface, which roughly translates into the amount of heat that would be generated by more than 300,000 average-sized nuclear plants.[7]

Scientists at the Goddard Institute for Space Studies in the United States have assembled temperature records from monitoring stations around the world that go back to 1866. According to their data, the global average temperature at the surface of the earth in the mid-1990s—15.27 degrees Celsius—was 0.6 degrees Celsius (1.1 degrees Fahrenheit) warmer than the average temperature recorded in the 1890s. (See Figure 1.) This makes the 1990s the warmest decade on record so far—despite a sharp cooling after the Mount Pinatubo volcano erupted in 1991. Although temperatures fluctuate naturally from year to year and decade to decade, it is striking that all ten of the warmest years since record-keeping began have occurred since 1980. And 1995 was the warmest year of all.[8]

Some have argued that this apparent warming is contradicted by satellite measurements of atmospheric temperatures, which show a slightly different trend. However, the satellite data, though more complete than ground-based measurements, simply represent temperatures at higher levels in the atmosphere, where the greenhouse effect is not as strong. The depletion of the ozone layer appears to be cooling portions of the upper atmosphere, which is also thought to contribute to the discrepancy. Moreover, compared to

FIGURE 1
Global Average Surface Temperature, 1866–1995

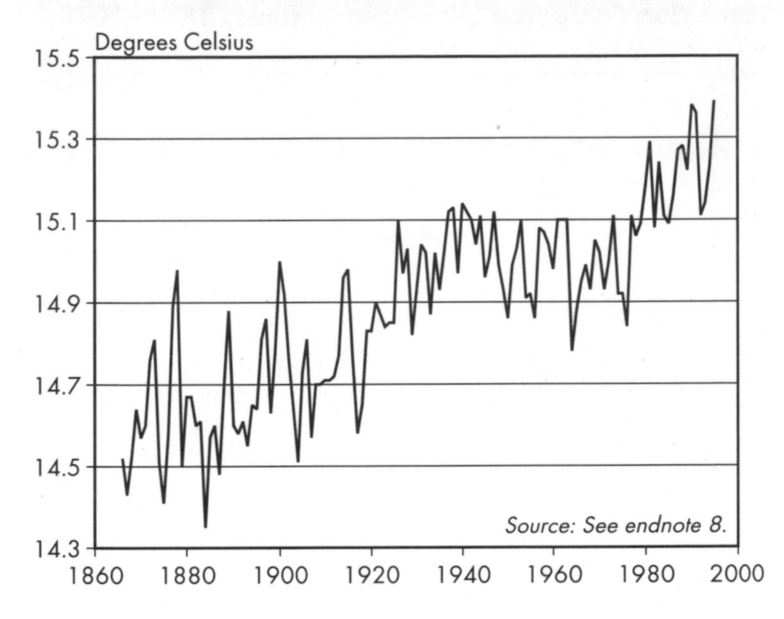

surface measurements, which have been compiled over 130 years, the satellite data cover only 17 years, too short a time span to establish any long-term trend. Over time, the satellite observations are expected to confirm the warming trend indicated in the surface temperature record. Meanwhile, another satellite-based instrument—a new orbiting radar gun—has detected a steady annual rise in sea level of 3 millimeters from 1992 to 1995, a trend consistent with the thermal expansion of water that occurs with warming.[9]

As early as the 1950s, scientists used crude "global circulation models" to simulate the atmosphere and climate, and these have become vastly more sophisticated in recent years. Designed and assembled by large teams of scientists and programmers, they take days to run on the fastest mainframe computers. The models must replicate many complicated feedback loops imitating the interaction of various climatic components. One such loop involves vaporization: higher temperatures tend to increase the amount of water

FIGURE 2

Temperature Projections with and without Sulfate Correction Compared with Observations, 1860–2040

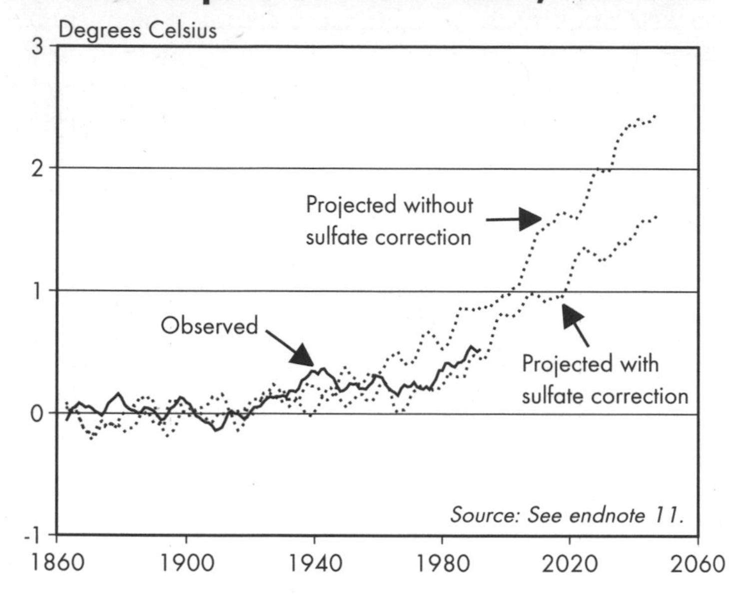

vapor in the atmosphere, which in turn causes additional warming. However, some of the water vapor turns into clouds, which can have a cooling effect, a factor that must also be replicated in the models. New "coupled" models of the atmosphere and oceans allow scientists to simulate the reciprocal actions of these two complex systems. Altogether, the models now do a remarkably good job of replicating past climate trends as well as regular climatic phenomena such as the change of seasons.[10]

One of the most important recent improvements in global circulation models is the incorporation of the effect of sulfate aerosols. These short-lived gases, which also come from fossil fuel combustion and cause acid rain, turn out to produce a cooling effect by forming a haze that reflects sunlight back into space. In 1994, the Hadley Centre for Climate Prediction and Research at the Meteorological Office in Bracknell, England, incorporated the aerosols in its model, and succeeded in mimicking global temperature records over the past century far more closely than earlier models did. As indicated by the Hadley model, the earth is in fact warming more rapidly in the southern hemisphere than in the north, where part of the warming effect of greenhouse gases is negated by this aerosol haze.[11] (See Figure 2.)

Analyses of regional climate shifts have helped confirm the overall trend. Since greenhouse warming is an uneven phenomenon—the poles should warm more rapidly than equatorial regions, and continents more rapidly than areas of deeply circulating ocean—scientists believe that certain localized "fingerprints" of climate change may be identifiable early on. Antarctica, for example, is warming rapidly, as climate models projected. This was vividly demonstrated in early 1995, when a 2,700-square-kilometer chunk of the Larsen Ice Shelf in Antarctica—as big as the U.S. state of Rhode Island—collapsed into the South Atlantic. Antarctic scientists now estimate that five of nine ice shelves attached to the Antarctic Peninsula have disintegrated in the past 50 years. [12]

Another region that is warming particularly rapidly is Siberia, which is now 3 degrees Celsius (5 degrees Fahrenheit) warmer than at any time since the Middle Ages. Northern Europe, meanwhile, experienced a string of warm winters and severe winter storms in the early 1990s, causing extensive flood damage. Related to this warming is a retreat of Alpine glaciers, exposing things that have been buried for thousands of years—including the famous "ice man," a 5,000-year-old frozen corpse found by hikers in 1992. He had tried to cross the Alps three millennia before the Roman Empire—the last time the Alps were covered with so little ice.[13]

The biological world is also showing signs of climate change. Oceanographers at the Hopkins Institute in Monterey, California, which has been tracking undersea life for 60 years, say that marine snails and other mollusks normally found in warm waters are expanding their ranges north along the Pacific Coast. Researchers have also linked the 80 percent decline in zooplankton off the southern California coast to a minor increase in the local surface water temperature over the last four decades. In northern Finland, pine trees are moving into the tundra at a rate of 40 meters per year in apparent response to warmer temperatures, while the *Aedes aegypti* mosquito—a vector for dengue fever and yellow fever—has been found at over 2,000 meters above sea level in parts of South America—1,000 meters higher than its previous range.[14]

Thomas Karl, Senior Scientist at the National Climatic Data Center of the U.S. National Oceanic and Atmospheric Administration, reported in 1995 that an analysis of several decades of U.S. weather data indicates that temperature and precipitation extremes have become more common in recent decades, a trend that is "consistent with the general trends expected from a greenhouse-enhanced atmosphere."[15]

Data on the timing of the seasons may provide another barometer of climate change. In early 1995, David J. Thomson, an expert in mathematical analysis of complex trends at AT&T's Bell Labs, published an article in *Science* on the changes in seasonal cycles in Europe, incorporating data from as far back as 13th-century church records. His data indicate that after centuries of relative stability, the timing of the seasons began to shift in the 1940s, showing a close correlation with the rise in concentrations of greenhouse gases.[16]

The cumulative weight of these indicators convinced the IPCC to conclude in its 1995 report that recent changes in global climate trends are almost certainly related to the rapid buildup of greenhouse gases in recent decades. Klaus Hasselmann, director of the Max Planck Institute for Meteorology in Hamburg, Germany, and one of the world's leading climatologists, has estimated a 95 percent chance that the observed changes in the earth's climate exceed the range of natural variability and therefore are linked to the rise in greenhouse gases. According to him, the smoking gun has been found.[17]

These conclusions have gained wide acceptance among scientists and policymakers, but have failed to persuade the so-called climate skeptics, who argue that computer-generated models are too flimsy to be predictive, that the temperature record shows a slower rate of warming than the models suggest, or that "negative feedbacks" will protect us from climate change. Scientists such as Patrick Michaels of the University of Virginia, Richard Lindzen of the Massachusetts Institute of Technology, and Robert Balling of Arizona State University have shifted frequently from one argument to another in an effort to minimize the risks of climate change. Many of their arguments are hardly reflected in the scientific literature, and their credibility is further undermined by the fact that they have been heavily funded by the oil and coal industries, and even by the government of Kuwait. Although their arguments are not regarded as plausible by most scientists, the skeptics have received considerable media coverage, partly because of an attempt to create "balance," which has confused the public about the state of climate science and helped slow international climate negotiations.[18]

Recent changes in global climate trends are almost certainly related to the rapid buildup of greenhouse gases.

Most scientists are now focusing on efforts to better anticipate future climate patterns by using the latest global circulation models. Responding to improvements in these models, such as the inclusion of the cooling effect of sulfate aerosols, and a better understanding of the global carbon cycle, the IPCC has revised its 1990 projections of future climate change slightly downward. The 1995 assessment now projects a rise of 1.0-3.5 degrees Celsius (1.8-6.3 degrees Fahrenheit) in the global average temperature between 1990 and 2100. This projection assumes, however, that sulfate pollution will continue to increase, an assumption that is belied by the fact that it has leveled off since 1990—the base year for the climate projections. During the next two decades, sharp sulfur emission reductions in Europe and North America seem likely to at least offset the growth projected in the developing world. If so, climate change may occur more rapidly than the modelers predict.[19]

Still, according to the IPCC, even at the lower end of the projections, the global climate will change faster in the next few decades than at any time since civilization began, as carbon dioxide concentrations reach levels the world has not seen for thousands of years.[20] (See Figure 3.)

The dramatic nature of these potential changes is seen in the fact that the global average temperature was just 3-5 degrees Celsius (5-9 degrees Fahrenheit) cooler during the last ice age, which ended 12,000 years ago, than it is today. During that period, glaciers covered most of Europe, and what is now New York City was buried under 100 meters of ice. If the global temperature were to warm by a similar amount, the changes—in the opposite direction—might be equally dramatic. The last ice age took several thousand years to roll back, yet greenhouse warming could cause equivalent change in just two centuries.[21]

Uncertainty about the rate and magnitude of climate change is caused by the complexity of the climate system and its many feedback loops. A lot depends on how the atmosphere, oceans, ice sheets, clouds, and biosphere respond to changing climatic conditions. One potentially dangerous area of uncertainty centers on the fact that the oceans and biosphere contain a substantial reservoir of carbon. Indeed, of the roughly 7 billion tons of carbon released from fossil fuel burning and land clearing each year, only about 3 billion tons remain in the atmosphere. Large quantities of carbon are apparently being absorbed by oceans and forests. The recent discovery that as much as half of this carbon may be going into the Northern Hemisphere's forests was surprise to many scientists, who believe this phenomenon may result in part from higher temperatures and the fertilizing effect of more CO_2 in the atmosphere.[22]

FIGURE 3

Local Temperature and Atmospheric Carbon Dioxide Concentrations at Vostok, Antarctica, 158,000 B.C. to Present

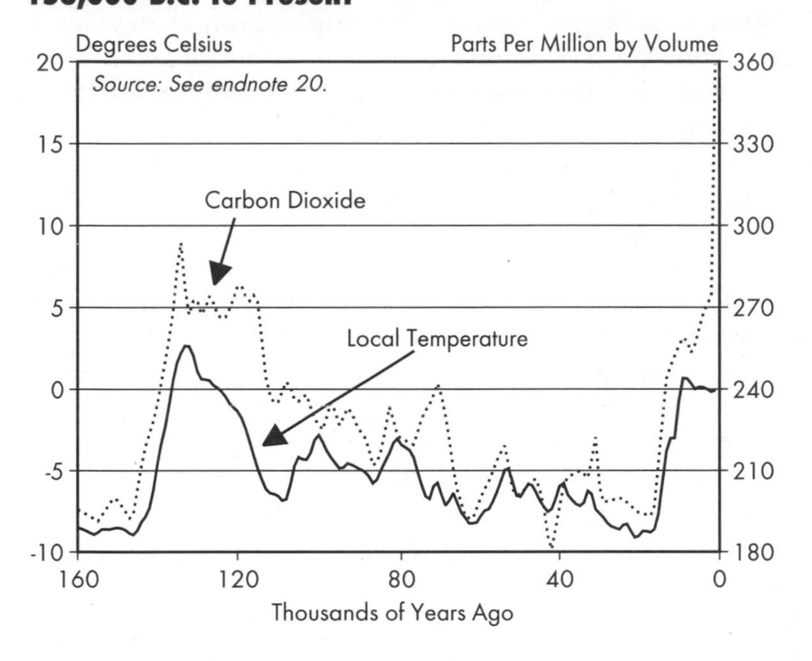

This absorption of carbon may be protecting the planet from the full effects of fossil fuel combustion, but it also foreshadows dangers. Some analysts are concerned about how resilient the terrestrial carbon sink is. Warmer temperatures could damage the health of forests, causing them to die and give up the carbon stored in their biomass and in surrounding soils. Extensive loss of northern boreal forests could set free tens of billions of tons of additional carbon, which would then accelerate the rise in temperatures. The warming of the tundra could also release large quantities of methane. Such effects could cause the rate of climate change to approach or exceed the upper range of the IPCC's projections.[23]

The ability of the oceans to absorb carbon may also be at risk. A report in *Nature* in August 1995 suggests that the oceans may be losing fixed nitrogen, an essential fertilizer that allows phytoplankton to grow, assimilating carbon that falls into the deep ocean when the phytoplankton die. If the oceans lose nitrogen as they warm, they will tend to absorb less carbon, thus boosting the rate of CO_2 buildup in the atmosphere. Although the net effect of these feedbacks is uncertain, they do suggest—contrary to the arguments of some skeptics—that uncertainty cuts both ways.[24]

For policymakers, this uncertainty presents a major challenge: how to respond to a problem whose precise dimensions are unclear. Yet in many other fields, governments and individuals frequently take action in the face of similar, or even greater, lack of certitude. Homeowner investments in insurance and government investments in military armament are obvious examples of people taking action to reduce risks despite large unknowns. A concept known as the "precautionary principle" suggests that with a problem such as global climate change it would be illogical to wait for absolute certainty in the form of a catastrophic climatic event before slowing greenhouse gas emissions.

A wait-and-see approach would likely commit the world to even faster climate change that would take centuries to reverse—extending the unplanned and unregulated experiment with the atmosphere that is already underway. As most governments implicitly acknowledged by signing the convention itself, the world cannot afford to continue ignoring the climate problem.

Climate Shocks

Climate change is likely to be an erratic, unpredictable process, marked by sudden discontinuities, according to recent studies. In its 1995 assessment, the IPCC concluded that "the incidence of floods, droughts, fires and heat outbreaks is expected to increase in some regions" as temperatures rise. Some changes could occur suddenly, affecting one region profoundly while leaving others undisturbed.[25]

In an age when many people live in air-conditioned homes and eat fresh food grown thousands of kilometers away, it is easy to ignore our dependence on the climate. Most cities are located near adequate water supplies, and their nutritional and material needs are met via agricultural, forestry, and fishery systems that require particular temperature, rainfall, and humidity levels. We can cope with isolated droughts, storms, or floods by bringing in relief supplies, but the widespread, simultaneous disruptions that may result from greenhouse warming would be unmanageable.

"The incidence of floods, droughts, fires and heat outbreaks is expected to increase in some regions." (IPCC, 1995)

Although a warmer climate will tend to boost both evaporation and precipitation, atmospheric models suggest that the regional effects would be uneven; unfortunately, many of today's humid regions are likely to get even more rain, while many continental regions become drier, intensifying existing conditions. With many regions—including the Middle East, northern China, northern India, and the western United States—already suffering from water scarcity, any further reduction of water supplies in these areas would be serious. Chronic water shortages currently plague more than a third of the world's population, according to the World Bank, impinging on agricultural output and economic development in some 80 nations.[26]

Global circulation models indicate that climate change could increase agricultural productivity in some northern regions as a result of longer growing seasons and increased precipitation. But most of the areas that might benefit, including northern Canada and Siberia, lack the rich soils now found in areas such as the midwestern United States and Ukraine, and so would be unable to deliver the bumper crops produced in today's "grain belts." Also, higher summer temperatures would impede pollination and boost evaporation, reducing soil moisture. Higher temperatures would also allow insects and diseases to proliferate, which would further reduce yields. Potential disruptions are shown by the fact that between 1988 and 1995, three U.S. grain harvests were diminished by heat and drought, contributing to a rapid drawdown in global grain reserves.[27]

South and Southeast Asia, tropical Latin America, and sub-Saharan Africa are among the regions that could see a decline in their subsistence crop harvests, raising the risk of famine in these already vulnerable areas. Nor would industrial countries in the Northern Hemisphere be immune to the adverse agricultural effects of climate change. A recent study by the European Commission concluded that climate change could reduce agricultural production in Europe, particularly in southern countries, where crop productivity already lags behind levels found in the North.[28]

Grasslands are particularly vulnerable to climate change since they are relatively short of moisture to begin with. Combined with other stresses such as overgrazing, even a minor reduction in rainfall or increase in evaporation could turn some of these grasslands into deserts, a process that is difficult to reverse. Already, grasslands are rapidly disappearing in the African Sahel and other regions.[29]

According to a study cited in the 1995 IPCC report, climate change could also reduce the flow of many rivers. The Indus River, in Pakistan, which waters the world's largest irrigation system, could see its average flow decline 43 percent by the end of the next century. Similarly, the Niger River, which provides water for much of northwest Africa, is projected to have its flow fall 31 percent during the same period. Overall, the IPCC finds, "there may be significant adverse consequences for food security in some regions of the world." At a time when growing populations and rising incomes are driving food demand upward, the world can ill afford such losses.[30]

Scientists also believe that drastic climate change is likely to reshape the natural world. Although climate has been a major driver of biological evolution since the beginning of life, occasionally wiping out thousands of species and entire ecosystems, the rate of change now projected far exceeds natural rates, and could cause ecological devastation in the decades ahead. In fact, minor shifts in seasonal temperature, even if well within the physiological limits of an organism's tolerance, can eliminate its food supply, which in turn can cause an entire ecosystem to unravel. For example, warming water temperatures seriously threaten the health of tropical coral reef ecosystems, which not only provide habitat for many fish and plant species, but also are valuable for medical research, fishing, and tourism.[31]

Higher temperatures also affect species' migration and spawning patterns. As temperatures warm, many plants, fish, insects, and bacteria may move into new areas, eradicating indigenous species. It is hard to know how severe these effects will be, but when combined with other stresses, including land clearing and local pollution, climate change could have catastrophic consequences for some ecosystems.[32]

Warming ocean temperatures, altered stream flows, and saltwater encroachment are projected to result in an 8 percent decline in the annual world fish catch by the year 2100. Recent studies find a threat to the Pacific Northwest salmon supply, where most of the fish could either perish or be forced to migrate into colder waters. The loss of these species could devastate the local communities that depend on them and reduce world protein supplies as well.[33]

Studies also indicate that half the world's coastal wetlands could be inundated in the next century. Regions that could be hit particularly hard include the coasts of West Africa, Australia, the Mediterranean, and large areas of East Asia, especially the Philippine archipelago and Papua New Guinea. The loss of wetlands is likely to damage fisheries as their spawning and nursery grounds are engulfed.[34]

Most forests are adapted to particular moisture and temperature conditions, and probably could not extend their range fast enough to keep up with human-induced climate change. The IPCC estimates that temperate forest zones could shift toward the poles by 160-640 kilometers over the next century—compared to a natural rate that ranges from less than 1 to 200 kilometers in the same time span. According to Steven Hamburg, an ecologist at Brown University, a staggering one-third of the earth's forests could be forced to "move" as a result of the effective doubling of CO_2 concentrations projected by 2100. Moreover, weather-related stress would likely make many forests vulnerable to insects, disease, and wildfires. Massive, sudden loss of northern forests is possible, which could spell disaster for the wood-products and tourism industries, as well as native species of plants and animals. The IPCC projects that "entire forest types may disappear" over the next century, including half the world's tropical dry forests.[35]

The 1995 assessment by the IPCC also concludes that "projected changes in climate are likely to result in a wide range of human health impacts, most of them adverse, and many of which would reduce life expectancy." For example, millions of people, including the sick and elderly, would suffer from temperature extremes like the record summer heat waves that killed hundreds in Chicago and other U.S. cities in 1995.[36]

Particularly alarming is the likelihood, spelled out in recent studies, that the animals and insects that carry infectious diseases such as malaria, cholera, and dengue fever could thrive in the warm, moist conditions likely to be more widespread in the coming decades. A Dutch study projects that a 3-degree rise in global temperatures could double the incidence of diseases carried by mosquitoes in tropical regions and increase them tenfold in temperate areas. Similarly, scientists believe that higher ocean temperatures could trigger algae blooms, leading to cholera epidemics; such algae are thought to be responsible for the 1991 cholera outbreaks in Southeast Asia and South America. At a time when the incidence of many infectious diseases is again on the rise due to human actions, climate change could complicate an already unmanageable public health challenge.[37]

Another aspect of climate change now being evaluated is its effect on the frequency and severity of storms. A scientific assessment for a German insurance company, Munich Re, notes: "A warmer atmosphere and warmer seas result in greater exchange of energy and add momentum to the vertical exchange processes so crucial to the development of tropical cyclones, tornadoes, thunderstorms, and hailstorms." Such links have not been definitively confirmed, but recent weather trends suggest that severe storms may already have become more common. A study by the U.S. National Oceanic and Atmospheric Administration, for example, reports a "steady increase in precipitation derived from extreme one-day precipitation events" in the United States in recent decades. And in Europe, severe winter storms have become more frequent, causing more than $10 billion in damage in 1990 alone.[38]

A 3-degree rise in global temperatures could double the incidence of diseases carried by mosquitoes.

Hurricanes, cyclones, and typhoons—as they are variously called in different parts of the world—are the most widely destructive and life threatening of natural disasters. These large, swirling storms originate in warm tropical waters such as the Caribbean Sea and the South Pacific and Indian oceans. They have destructive winds of 120-320 kilometers per hour, and are accompanied by heavy rain and storm surges that inundate low-lying areas. Meteorologist Kerry Emanual of the Massachusetts Institute of Technology estimates that the 3-4 degree Celsius rise in sea temperatures projected by atmospheric models could increase the destructive potential of hurricanes by 50 percent and cause sustained storm winds as high as 350 kilometers (220 miles) per hour.[39]

Such a warming could lengthen the current hurricane season in North America by two months or more, and allow the storms to move further north, striking major urban areas such as New York City before petering out, according to Donald Friedman, former director of the Natural Hazards Research Program for the Travelers Insurance Company. Other scientists dispute these numbers, however, noting that tropical storms require a complex brew of forces, and that some features of a warmer climate could make it more difficult for hurricanes to form.[40]

During the past five years, the world has experienced unprecedented damage from weather-related disasters, possibly a preview of things to come. (See Table 1.) In May 1991, for example, a cyclone with winds of 270 kilometers per hour hit Bangladesh, flooding vast areas of the country's coastal plain. An estimated 140,000 people were killed, and more than a million homes were damaged or destroyed. Financial losses were put at $3 billion—more than 10 percent of Bangladesh's annual gross national product (GNP). And in 1995, East Asia was hit by heavy rains and floods that caused damage worth $6.7 billion in China and $15 billion in North Korea, where crops were so devastated that emergency food supplies had to be brought in.[41]

After two decades of relative calm, the southeastern United States was struck by a number of serious storms in the 1990s, particularly in 1995, which had the most active Atlantic hurricane season since the 1930s. Although warning systems limited the loss of life, economic damage was extensive because of burgeoning coastal development. The most severe storm was Hurricane Andrew—the third most powerful hurricane to make landfall in the United States in the 20th century—which hit south Florida in August 1992 with sustained winds of 235 kilometers per hour. Andrew virtually flattened 430 square kilometers of Dade County in Florida, destroying 85,000 homes and leaving almost 300,000 people homeless. Losses were estimated at $30 billion—equivalent to the combined losses of the three most costly previous U.S. storms. Robert Sheets, then director of the National Hurricane Center, estimated that if Andrew had moved just 30 kilometers further north, it would have

TABLE 1

Weather-related Disasters with Damages over Three Billion Dollars, 1990-1995

Disaster	Location	Year	Deaths	Estimated Damages (billion dollars)
Windstorm Daria	Europe	1990	-	4.6
Windstorm Vivian	Europe	1990	-	3.2
Cyclone	Bangladesh	1991	140,000	3.0
Flood	China	1991	3,074	15.0
Typhoon Mireille	Japan	1991	62	6.0
Hurricane Andrew	N. America	1992	74	30.0
Cyclone Iniki	N. America	1992	4	3.0
Winter storm	N. America	1993	246	5.0
Mississippi floods	N. America	1993	41	12.0
Winter storms	N. America	1994	170	4.0
Spring floods	China	1994	1,846	7.8
Flood	Italy	1994	64	9.3
Winter floods	Europe	1995	28	3.5
Flood	China	1995	1,390	6.7
Storm, flood	N. Korea	1995	68	15.0
Hurricane Opal	N. America	1995	28	3.0

Source: See endnote 41.

caused damage valued at $100 billion—principally in Miami and New Orleans, which would have been under 6 meters of water.[42]

Although hurricane severity is not definitively linked to climate warming, it is clear that hurricane losses could be exacerbated by another feature of a warming world: rising seas. Water expands as it warms, and higher temperatures also tend to melt the glacial ice found near the world's poles. During the past century, sea levels have already risen 20-40 centimeters, a range that is caused by regional differences such as ocean currents and the fact that some land areas are subsiding due to freshwater withdrawals and other effects. IPCC scientists believe that by 2100, sea levels will rise another 15-95 centimeters above current levels. And even if greenhouse gas concentrations in the atmosphere are stabilized, because of time lags in the climate system, sea levels are expected to continue rising for several decades beyond the point at which emissions level off.[43]

Rising sea levels exacerbate the threat that climate change presents to coastal communities and the estuaries and aquifers on which they depend. According to the IPCC, rising seas could flood many deltas, and make portions of some cities uninhabitable. Its mid-range projections indicate that most of the beaches on the East Coast of the United States may disappear in the next 25 years. Estimates for the Netherlands suggest that the country would have to spend $3.5 trillion—ten times its annual GNP—to build up its dikes sufficiently to fend off the rising North Sea.[44]

A study by the Asian Development Bank found that Bangladesh, India, Malaysia, the Philippines, Sri Lanka, and Vietnam would be particularly hurt. The ocean around Jakarta could rise 1 meter by 2070, submerging a large portion of the city. In Vietnam, rising seas could inundate much of the Red River and Mekong deltas, cutting production of rice, the country's principal crop. Bangladesh and China are each projected to have 70 million people displaced by rising seas during the next century—at a time when they are struggling to cope with burgeoning populations. In Africa, some 16 percent of Egypt's population could be displaced by late in the next century, and low-lying areas of Mozambique are also at risk. Climate change could make environmental refugees as common as political refugees are today.[45]

Small island countries are particularly vulnerable to the combined effects of rising sea levels and stronger storms. Ieremia Tabai, former president of Kiribati, observes: "If the greenhouse effect raises sea levels by one meter it will virtually do away with Kiribati....In 50 or 60 years, my country will not be here." The same is true of at least a dozen other small island nations as well as various states and territories in the Caribbean and tropical Pacific. The Maldives, the Marshall Islands, Tokelau, and Tuvalu are particularly vulnerable.[46]

Some industries are also likely to be hit hard by climate change. Among those on the front lines are agriculture, fishing, forestry, and tourism. Total economic damage to the global agriculture industry alone is projected to amount to $40 billion annually by the end of next century. Although, few of these industries have thoroughly examined the potential for damage, some wary investors are beginning to assess the dangers. Sven Hansen, a vice president at Union Bank of Switzerland, notes that "some of our clients are under major threat from climate change . . . in my opinion, it is the single most important environmental problem for the world today."[47]

The insurance industry, which pays out claims to vulnerable industries, may feel the pain first. Since 1990, insurers have paid out $57 billion for weather-related losses worldwide, compared with $17 billion for the entire decade of the 1980s. (See Figure 4.) Although it is not certain the losses were caused by climate change, H.R. Kaufmann, General Manager of Swiss Re, one of Europe's largest insurers, believes that "there is a significant body of scientific evidence indicating that last year's record insured loss from natural catastrophes was not a random occurrence....Failure to act would leave the industry and its policyholders vulnerable to truly disastrous consequences." Hurricane Andrew alone wiped out seven U.S. insurance companies.[48]

The dilemma for insurers is that their rates and coverage policies are based on the law of averages. In the case of weather-related coverage, they assess past trends and assume that the frequency of catastrophes will stay the same. But climate change could render those calculations useless. In response, many companies are reducing their exposure in coastal and island real estate, wildfire-prone regions, and

FIGURE 4

Economic Losses from Weather-related Natural Disasters Worldwide, 1980–95

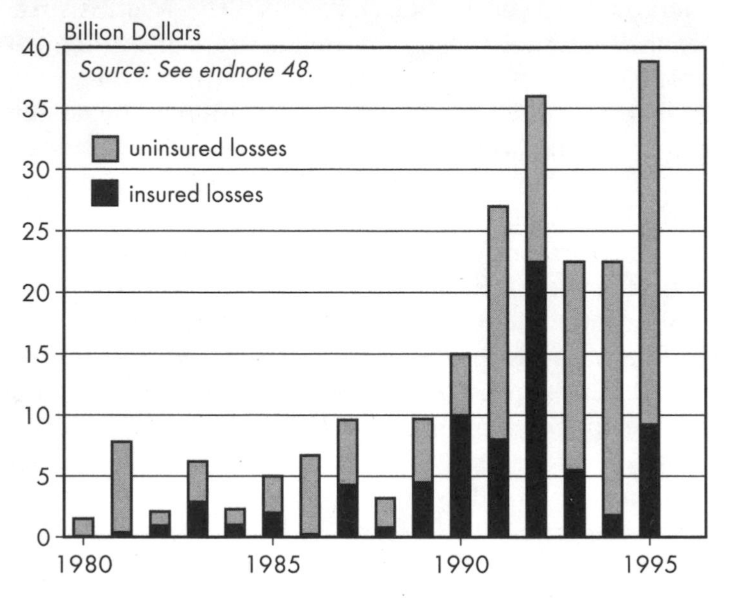

valleys vulnerable to flooding. But if the industry abandons its riskier policies, either governments will have to step in as insurers of last resort or society will lose a vital buffer against weather-related disasters.[49]

Economists have assessed the potential financial consequences of global climate change, though such estimates are rough at best due to enormous uncertainty about the specific regional effects of climate change. In one of the most comprehensive studies, William Cline, an economist at the Institute for International Economics in Washington, D.C., concluded that the damage to the U.S. economy from a doubling of effective CO_2 concentrations could reach $53 billion per year—or 1 percent of the current annual GNP. Other estimates range as high as $120 million annually. But since the world suffered a record $38 billion in financial losses from weather-related disasters in 1995 alone, the uncertainty of the estimates is clear.[50]

For developing countries, the economic effects could be catastrophic. The 1995 floods in North Korea, for example, brought the country's economy to the verge of collapse. For nations struggling with a range of other environmental problems and already short of capital for development—which includes most countries in Africa and South Asia—the proliferation of such disasters would severely undermine economic prospects.[51]

Reducing Greenhouse Gas Emissions

Under the Framework Convention on Climate Change, industrial countries committed themselves to the aim of holding their emissions of greenhouse gases at or below the 1990 level in the year 2000. This goal, which some countries interpret as legally binding while others do not, applies to industrial countries that belong to the Organisation for Economic Co-operation and Development (OECD), and to the nations of Eastern Europe and the former Soviet Union as well. Though it is intended only as a downpayment on the eventual need to sharply cut emissions, this commitment is currently being missed by roughly half of the countries to which it applies. Meanwhile, emissions soared in developing countries in the first half of the 1990s, driving the global figures upward.[52]

Between 1990 and 1995, annual fossil-fuel-related emissions of carbon, which produce carbon dioxide, the leading greenhouse gas, rose by 113 million tons, reaching 6 billion tons in 1995. They would have risen an additional 400-500 million tons if not for the economic collapse of central and eastern Europe, which temporarily drove down emissions in that region. (See Figure 5.) Roughly 1.6 billion tons of additional carbon are released annually from land clearing, primarily in tropical regions such as the Amazon Basin. Emissions of CFCs are falling sharply as a result of efforts to protect the ozone layer, and trends for the other greenhouse gases are difficult to track. Some countries report success in bringing down emissions of methane and nitrous oxides.[53]

Greenhouse gas emissions are nearly ubiquitous in today's economies. Whether clearing a field, cultivating a rice paddy, cooking a meal, turning on a light, or driving a car, we are all contributing to climate change. Some of these activities contribute more than others, however. Industrial processes generate about 40 percent of total emissions; transportation comprises roughly 20 percent, and residential and commercial buildings add another 20 percent. The remaining 20 percent comes from a broad range of other activities, particularly land clearing, agriculture, and forestry. The fastest increases in emissions are occurring in transportation and generation of electric power, both of which are expanding particularly rapidly in developing countries.[54]

Greenhouse gas emission levels vary widely among nations. Per capita emissions of carbon dioxide from fossil fuels range from 5.3 tons in the United States to 2.4 tons in Japan and 0.2 tons in India. (See Table 2.) This more than 20-fold range in emission rates reflects many differences, including levels of industrial development and personal incomes. But one striking feature of the world carbon budget is the wide range of emissions even among countries at similar levels of economic development: per capita emissions are 75 percent higher in China than they are in Brazil, for instance, while in the United States they are 120 percent higher than in Japan. Such differences reflect variations in the energy efficiency levels of individual countries and in the fuel mixes each nation relies on.[55]

Figures on the amount of carbon from fossil fuels emitted per million dollars of economic output—a measure of the carbon efficiency of economies—also show great disparities. Among the least carbon-efficient economies are Kazakhstan at 1,250 tons of carbon per million dollars of GNP, South Africa at 680 tons, and Russia at 590 tons. The United States, by contrast, emits 210 tons of carbon per mil-

FIGURE 5

World Carbon Emissions from Fossil Fuel Burning, by Economic Region, 1950–94

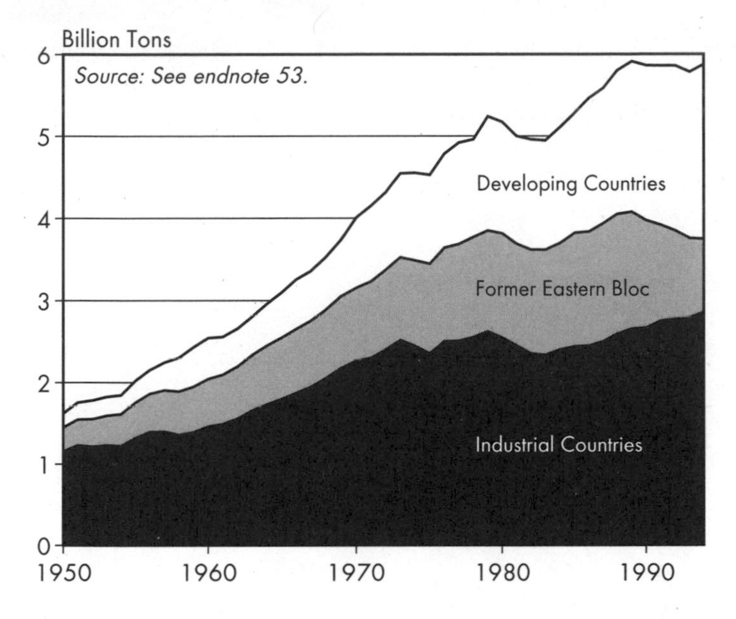

lion dollars of GNP, while Japan is responsible for only 110 tons. Developing-country emissions range from 330 tons of carbon per million dollars of GNP in China to 160 tons in India and 70 tons in Brazil. These data reflect many factors, including inefficient heavy industry in the former Soviet Union, extensive reliance on automobiles in the United States, high levels of energy efficiency in Japan, heavy use of coal in China and South Africa, and extensive use of non-carbon-emitting hydropower and biomass energy in Brazil.[56]

Under the terms of the climate convention, all countries must prepare a full inventory of greenhouse gas emissions as well as a national climate plan. However, only industrial countries are required to adopt national emissions reduction targets in line with the overall goal of the convention, and to implement action plans to achieve them. The national climate plans developed so far vary widely in character, but most are having only a limited effect on emis-

TABLE 2

Carbon Emissions from Fossil Fuel Burning in the Top 20 Emitting Nations, 1994

Rank	Country	Total Emissions	Emissions per Person	Emissions per Dollar GNP	Emissions Growth 1990-94
		(million tons)	(tons)	(tons/million $)	(percent)
1.	U.S.	1,371	5.26	210	4.4
2.	China	835	0.71	330	13.0
3.	Russia	455	3.08	590	-24.1
4.	Japan	299	2.39	110	0.1
5.	Germany	234	2.89	140	-9.9
6.	India	222	0.24	160	23.5
7.	U.K.	153	2.62	150	-0.3
8.	Ukraine	125	2.43	600	-43.5
9.	Canada	116	3.97	200	5.3
10.	Italy	104	1.81	110	0.8
11.	France	90	1.56	80	-3.2
12.	Poland	89	2.31	460	-4.5
13.	S. Korea	88	1.98	200	43.7
14.	Mexico	88	0.96	140	7.1
15.	S. Africa	85	2.07	680	9.1
16.	Kazakhstan*	81	4.71	1,250	n/a
17.	Australia	75	4.19	230	4.2
18.	N. Korea*	67	2.90	960	n/a
19.	Iran*	62	1.09	270	n/a
20.	Brazil	60	0.39	70	15.8

*Latest data available for these countries is from 1992.

Note: GNP data adjusted for puchasing power parity (PPP) is from 1993.
Source: See endnote 55.

sion trends.[57]

One country that is missing its national emissions target by a large margin is the world's leading carbon emitter—the United States. President Clinton's Climate Change Action Plan, launched at a White House ceremony in 1993, includes 50 measures intended to promote energy efficiency, commercialize renewable energy technologies, and encourage tree planting. Two-thirds of these measures are voluntary—consisting of relatively weak public-private partnerships. Still, some of these efforts, such as the "Green Lights" program started by the Environmental Protection Agency in the 1980s, have made considerable progress, encouraging manufacturers to produce a new generation of refrigerators, personal computers, and other devices that use a fraction as much energy as traditional models do. Another program, the "Climate Challenge"—a voluntary initiative involving electric utility companies—has accomplished little so far. Still other important elements of U.S. climate policy actually predate the plan itself; tighter efficiency standards for home appliances and lighting, for example, were included in the Energy Policy Act passed by the U.S. Congress in 1992.[58]

The U.S. climate plan does not include any new taxes on energy use or carbon dioxide emissions, and the country's fuel prices remain among the lowest in the world, encouraging heavy use of energy. Nor does the U.S. plan include automobile fuel economy standards, which were left out due to the opposition of the nation's powerful automakers. To make matters worse, the 1995 Congress sharply cut funding for the government's climate plan, weakened the appliance and lighting standards that had been enacted three years earlier, and then abolished the national speed limit of 55 miles per hour. Since a car being driven at just 65 miles per hour emits 15-20 percent more carbon dioxide than one going 55, this change will lead to millions of tons of additional CO_2 being emitted each year.[59]

According to its own terms of reference, the U.S. plan is to hold carbon dioxide emissions to within 3 percent of the 1990 level in 2000, while reductions in emissions of other gases offset the increase. However, a 1996 projection by the U.S. Department of Energy indicates that without new policy initiatives, U.S carbon emissions in 2000 will exceed 1990 levels by a full 11 percent. This gap stems from the weakness of the U.S. plan as well as from congressional funding cuts. And to complete the dismal North American picture, Canada has also failed to turn strong government rhetoric on the climate problem into effective action, with the result that emissions there are growing at a similar pace.[60]

The record on emissions trends and climate policy in Europe is a mixed one, with some countries well on the way to meeting their commitment to hold emissions to the 1990 level in the year 2000 and others falling well short. Overall, according to a 1996 assessment by the European Commission, western Europe's carbon dioxide emissions in 2000 are likely to exceed 1990 levels—perhaps by as much as 5 percent—unless energy prices are raised or other new policy measures are introduced.[61]

Germany, Europe's leading producer of greenhouse gases, has established a far tougher emissions reduction target than is required under the convention. At the March 1995 Berlin meeting, Chancellor Helmut Kohl pledged a 25 percent reduction in carbon emissions from the 1990 level by 2005, a goal that reflects strong public concern about climate change in Germany, as well as the fact that the country is getting some easy emission reductions as the energy-intensive industries of eastern Germany are closed down or switched from coal to gas. By 1995, German carbon dioxide emissions were already more than 10 percent below the 1990 level.[62]

Germany's climate plan incorporates a number of solid measures, such as incentives for improving the energy efficiency of buildings and a new "electricity in-feed law" that grants generators of renewable electricity the right to sell power to utilities at a generous price of 17 pfennigs (11¢) per kilowatt-hour. As a result of this law, Germany installed more wind turbines in 1995 than any other country and is building hundreds of small power plants whose waste heat is used for district heating and other purposes.[63]

Yet German climate policy is plagued by contradictions. Coal is the country's traditional energy source, and the industry supports tens of thousands of well-paying jobs in the Ruhr Valley. German coal is far too expensive to be competitive on its own, but until recently electric utilities were required to buy it, subsidized by a "kohlepfennig" (coal penny) surcharge of 8.5 percent on consumers' electricity bills. This served as a sort of reverse carbon tax (at a rate of roughly $50 per ton), encouraging use of the most carbon-intensive fuel. In the past it was argued that removing the subsidy for domestic coal would simply lead to its being replaced by imported coal, but it now seems likely that natural gas and renewable energy could fill the breach. After German courts declared the surcharge unconstitutional in 1995, the government decided to replace it with a direct subsidy of $5.3 billion annually—which will have the perverse effect of making electricity cheaper.[64]

By 1995, German carbon dioxide emissions were already more than 10 percent below the 1990 level.

A similar contradiction can be found in Germany's transport sector, where CO_2 emissions have continued to rise as passenger traffic and truck transport have surged—particularly in the eastern states, where road transportation had been quite limited. German highways have no speed limits and the government still reimburses car commuters at a fixed rate per distance driven regardless of the availability of other means of transport—because of a law enacted to encourage car sales in the 1950s.[65]

One of the bright spots in German climate policy is its high gasoline taxes, increased by 0.23 DM per liter in 1994, in part for environmental reasons. Germany's total gasoline tax of about 1 DM per liter ($3 per gallon) is now equivalent to $1,200 per ton of carbon. For several years, Germany has agonized over the enactment of a broader CO_2 or energy tax. In early 1996, the government backed down from its latest energy tax initiative after auto and chemical company leaders interceded with the chancellor. As a result, new energy taxes are unlikely to be considered in Germany until after federal elections in 1998. Still, the government used the threat of such a tax to extract a commitment from the Federation of German Industries—a coalition of 19 industry associations—to voluntarily cut CO_2 emissions per unit of production 20 percent by the year 2005. This is a tough goal that will require substantial investments by many of these industries.[66]

Germany will easily meet the treaty's target for the year 2000, but without additional policy measures it is likely to fall well short of the ambitious 25 percent reduction pledged by the government. Environmental groups in Germany believe that the lack of measures to control emissions from transportation represents the biggest gap in the country's climate plan.[67]

The strongest climate policies to date are found in two small nations: Denmark and the Netherlands. Denmark's Energy 2000 plan aims at reducing carbon emissions to 20 percent below 1988 levels by 2005. Much of the reduction is expected to be achieved through a system of CO_2 and energy taxes, which were again raised in early 1996. Denmark has also invested heavily in wind power, biomass energy, and other alternatives.[68]

In the Netherlands, the National Environmental Policy Plan targets a 3 percent reduction in emissions by 2000. Much of the reduction is projected to flow from "voluntary" covenants between government and industry—agreements that are much stronger than those the U.S. government has devised. The Netherlands has also made a major commitment to capturing waste heat from power plants, to cover half the country's power generation by 2000. Major efforts are also being made to improve the efficiency of buildings and appliances. Gasoline and automobiles are heavily taxed, and a new energy tax was applied to private households and other small energy consumers in 1996. Some 10 percent of the nations's surface transportation budget goes to bicycle facilities, and subsidies to public transportation were recently raised to $5.7 billion annually.[69]

According to an independent assessment by the European Commission, these two countries are within reach of achieving the year-2000 emissions target laid out in the treaty. They have shown that it is possible to turn paper promises into concrete actions, and their example could well show the way for other nations. Of course, the fact that these smaller nations do not have large automobile or coal industries to contend with made it easier to adopt strong climate policies.[70]

Japan, which has a well-deserved reputation for its efficient use of energy as well as its commitment to new energy technologies, has had only limited success with its national climate program so far. It includes voluntary energy efficiency efforts, as well as new efficiency standards for appliances and industrial equipment, but these are not backed by new legislation aimed specifically at reducing emissions, or by new economic measures such as energy taxes. Consequently, Japanese electric utilities intend to increase coal burning, and automobile fuel economy is declining. Carbon emissions in Japan have leveled off since 1990, but mainly because the country's economy has stagnated. Emissions may be further restrained in coming years as many Japanese companies are pushed by the strong yen to move their factories offshore. Naturally, this will tend to reduce domestic use of fossil fuels.[71]

Ironically, the largest reductions in greenhouse gas emissions so far are in countries with virtually no climate plans. Owing to economic restructuring and a dramatic decline in the most energy-intensive industrial sectors, carbon emissions plummeted 20 percent in Russia, 27 percent in Poland, and 38 percent in Ukraine between 1986 and 1994; as a result, all these countries will easily meet the treaty's year-2000 target. Yet most of these countries still have extraordinarily inefficient energy systems and encourage wasteful use of energy via heavily subsidized energy prices and inadequate equipment and building standards. Energy reforms in the region, supported by outside funders, including the European Union and the European Bank for Reconstruction and Development, have focused heavily on the need to improve energy efficiency. If such programs succeed, the region's emissions could continue falling in the next decade, though growing automobile and truck traffic could offset some of the gains.[72]

Developing countries' greenhouse gas emissions, on the other hand, are rising rapidly, driven by pent up demand for transportation, refrigeration, lighting, and scores of other modern amenities. China's CO_2 emissions rose 13 percent between 1990 and 1994, Brazil's 16 percent, India's 24 percent, and South Korea's 44 percent. Growth in energy demand, which had been restrained by high oil prices, foreign debt, and economic stagnation in the 1980s, is now surging. These trends are projected to continue and will likely push global emissions up in the late 1990s. Still, in many Asian and Latin American countries, emissions are growing slower than their GNPs—thanks to the fact that their light industry and service sectors are developing faster than energy-intensive industries such as steel or chemicals.[73]

In China, for example, carbon emissions grew at 5 percent per year in the early 1990s, while economic growth averaged 10 percent annually. The government has made a concerted effort to improve the energy efficiency of Chinese industry. Yet China still has a 19th-century-style energy system—heavily dependent on coal, even for home heating and cooking—which has already made it the world's second largest emitter of carbon dioxide.[74]

Recognizing their growing energy needs, the climate convention requires developing countries to adopt national climate plans but not to meet any specific emission reduction targets yet. So far, few of the national plans have been completed, but it is hoped that they will help demonstrate that there are many cost-effective means of reducing greenhouse gas emissions. Some of these investments may ultimately be supported by the Global Environment Facility (GEF), an international fund authorized to help developing countries reduce their greenhouse gas emissions.[75]

The overall record on climate policy to date is not encouraging. At a March 1996 session to review progress in meeting the convention's goals, Michael Zammit Cutajar, the convention's executive secretary, reported that rather than stabilizing their emissions, most industrial countries are "heading in the opposite direction." If the trends in developing countries are taken into account, the picture is even darker. Overall, the biggest problem is transportation, where reliance on automobiles and trucks continues to grow. Burgeoning construction of inefficient buildings, and coal-fired power plants are also contributing to the growth of CO_2 emissions.[76]

China's CO_2 emissions rose 13 percent between 1990 and 1994, Brazil's 16 percent, India's 24 percent, and South Korea's 44 percent.

Unless additional policies are implemented, the International Energy Agency projects that global emissions of carbon dioxide from fossil fuels will exceed 1990 levels by 17 percent in 2000 and 49 percent by 2010, reaching nearly 9 billion tons of carbon annually. This represents a slowdown from the rates of growth that prevailed in the 1960s and 1970s, but emissions are still moving in the wrong direction.[77]

Stabilizing the Climate

The central goal of the Framework Convention on Climate Change is stated clearly and unambiguously in its second article: "The ultimate objective of this Convention...is to achieve...stabilization of greenhouse gas concentrations in the atmosphere at a level that would prevent dangerous anthropogenic interference in the climate system. Such a level should be achieved within a time-frame sufficient to allow ecosystems to adapt naturally to climate change, to ensure that food production is not threatened and to enable economic development to proceed in a sustainable manner."[78]

The ultimate test of the treaty, in other words, is its ability to protect both the natural world and the human economy from long-term damage stemming from rapid, uncontrolled climate change. As the treaty implicitly acknowledges, some additional climate change is inevitable during the next few decades, due to alterations in the composition of the atmosphere that have already occurred. Avoiding even more drastic climatic shifts is still within reach, but will require a major change in direction—not only slowing growth in emissions, but reducing them substantially.

As is often the case with complex international agreements, climate negotiators found it easier to lay out broad goals than to translate them into specific targets or policy measures. As the convention was assembled in 1992, negotiators dodged these broader questions and focused instead on short-term industrial country targets. But climate change is by its nature a global, long-term problem. The time has come to flesh out the convention's ultimate aims, translate them into national and international strategies, and then implement policies that can achieve them.

At the first Conference of the Parties in Berlin in 1995, treaty members agreed to draft a legally binding protocol with long-term objectives, to be adopted at the third

Conference of the Parties in 1997. The precise terms of that protocol have not yet been determined. To assist policy-makers, several recent studies have begun to explore the limits within which the energy economy will have to stay if the world is to be protected from overly rapid climate change. They show that it is the *rate* of warming as much as the absolute amount that will determine the scale of the human and ecological impact. While both people and natural systems may be able to adapt to slow change, they could be devastated by more rapid shifts, which are more likely to cause major disruptions.[79]

Many scientists believe that anything more than an increase of 0.1 degrees Celsius (0.18 degrees Fahrenheit) per decade—1.0 degrees Celsius (1.8 degrees Fahrenheit) over the next century—would present unacceptable risks to natural systems as well as the human economy. For example, ice core data extracted from Greenland indicate that when climate change occasionally has exceeded this rate in the past, northern hemisphere forests have been devastated.[80]

The global average temperature at the earth's surface has risen at more than twice the acceptable rate in recent decades, and so far shows no sign of letting up. Climate models indicate that to slow the rate of change to 0.1 degrees Celsius (0.2 degrees Fahrenheit) per decade, the total concentration of all greenhouse gases will have to be held to a maximum of less than 550 parts per million CO_2-equivalent at mid-century, compared to the current level of roughly 430 ppm CO_2-equivalent. This suggests that the concentration of carbon dioxide alone will have to level off at between 450 and 500 parts per million—compared to the current level of 360 ppm. (See Figure 6.)[81]

As climate negotiators have begun to look more closely at these crucial long-term cumulative goals, they have realized the stringency of what is required. (Most "business-as-usual" scenarios project that concentrations of CO_2 alone will exceed 600 ppm by late in the next century.) Dutch scientists have described what they call "safe emission corridors," future emission trajectories that would allow the

FIGURE 6

Atmospheric Carbon Dioxide Concentrations, 1900–95, with Projections to 2200

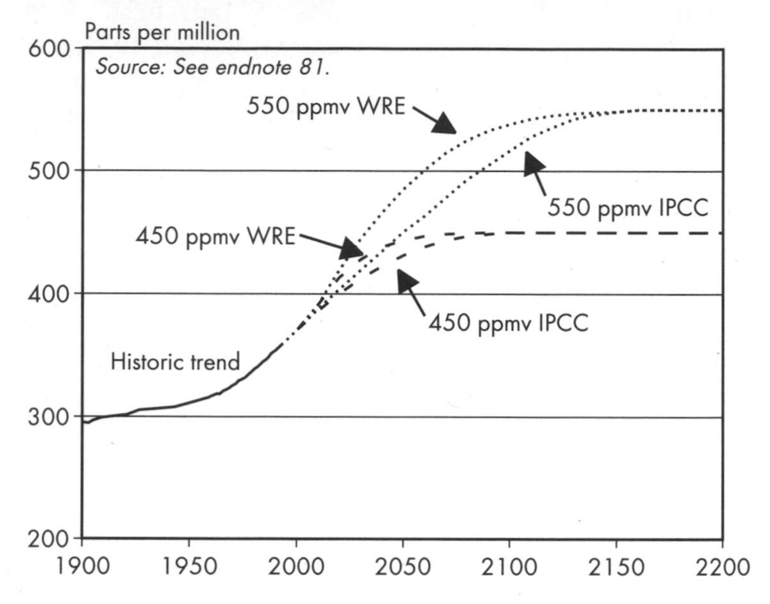

world's climate to stay within safe economic and ecological limits. As their models indicate, it is cumulative emissions over the next century that will determine at what level CO_2 concentrations can be stabilized. The 450 ppm CO_2 target means cutting emissions by more than half toward the end of the 21st century. (See Figure 7.) And if it turned out to be necessary to return concentrations to the *current* level, we would have to virtually eliminate carbon dioxide emissions by the middle of the same century.[82]

Although some increase in CO_2 emissions in the next decade can be allowed without exceeding the "safe emission corridor" defined by the 450 ppm ceiling, they cannot be permitted to rise as rapidly as governments are now projecting. And most of this "growth allowance" will have to go to developing countries whose emissions are currently low and whose economies are expanding rapidly. Industrial country emissions would not be able to grow at all in the short

FIGURE 7

World Carbon Emissions, 1900–95, with Projections to 2200

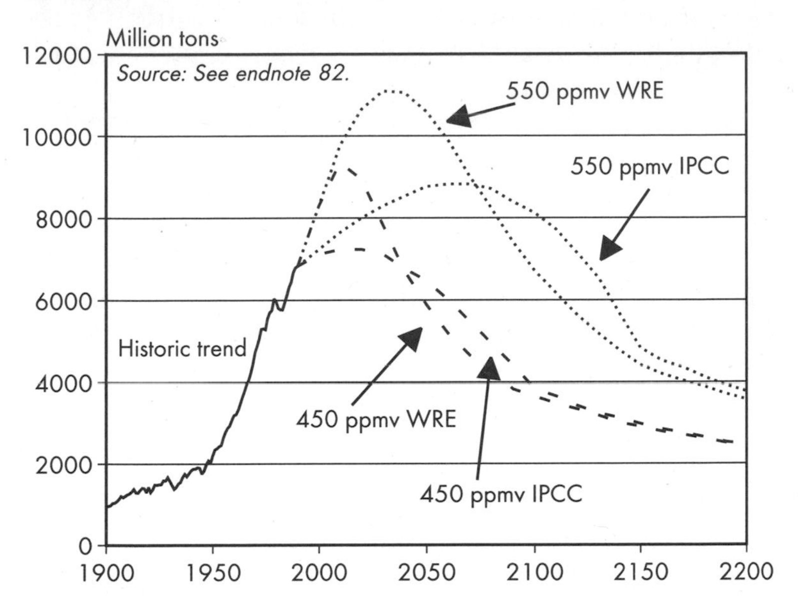

term—consistent with the Framework Convention's current provisions—and would have to begin falling steadily within a decade or so.[83]

The next few decades may be a particularly dangerous period. Continuing rapid increases in carbon dioxide and methane concentrations, combined with a leveling off or even a decline in sulfate aerosols—as a result of stringent sulfur emission standards enacted in industrial countries in the early 1990s—will tend to accelerate the pace of change. Consequently, action to limit emissions of the shorter-lived greenhouse gases—mainly methane and HCFCs—is essential during this period, along with strengthened efforts to limit CO_2 emissions.[84]

With the world's population projected to grow from 5.7 billion in 1996 to 8-10 billion in the next four decades, with the global economy likely to triple in size, and with people in developing countries looking forward to having modern amenities, substantially reducing carbon emissions will mean an end to the fossil fuel energy economy as we currently know it. We will have to shift to an energy system that uses fossil fuels not as its main source—currently they account for 80 percent of world energy supply—but as a minor supplementary one at best.[85]

To many energy companies, and to most ordinary citizens as well, a world without fossil fuels seems inconceivable. After all, these fuels powered the industrial revolution of the 19th century, fueled two world wars, and drove the unprecedented economic expansion of the postwar era. Yet the past few years have seen a growing optimism among analysts that the world can indeed find a practical, gradual path to a low-carbon energy system.[86]

The new optimism is based in part on careful analysis of historical trends, which indicate that in some respects the world is already headed toward a less carbon-intensive energy system—albeit at a slower pace than climate scientists believe is necessary. Jesse Ausubel, a professor at Rockefeller University, wrote in a 1995 article that "the global energy system has [for the last 100 years] been economizing on carbon." In fact, since human beings first burned wood tens of thousands of years ago, the world has moved steadily toward less carbon-intensive fuels. We have moved from wood to coal in the 19th century, to oil in the early 20th, and increasingly toward natural gas as we approach the 21st—lowering carbon intensity from nearly 1.1 tons of carbon per ton of oil equivalent in 1860 to 0.7 tC/tOE in 1990.[87]

The world energy system is not only becoming less carbon intensive, it is becoming more efficient in its use of energy. Since the turn of the century, the amount of energy required to produce a dollar of GNP in the United States has fallen by about 65 percent. This same index shows a reduction in energy intensity of 30 percent just since oil prices rose in the mid-1970s, a record that has been matched by most industrial countries. The scale of these achievements is demonstrated by the fact that if energy efficiency and carbon intensity in industrial countries had remained at the levels of 20 years ago, the world economy would currently be pumping nearly 8 billion tons of carbon into the atmosphere each year rather than 6 billion.[88]

Underlying technological and economic forces are likely to help extend the "decarbonization" of the global economy. Today, light manufacturing, services, and information technologies are growing faster than heavy industry, a development that will tend to slow growth in carbon dioxide emissions. Still, such shifts are unlikely by themselves to be sufficient. Simple calculations show that to meet the stabilization target of 450 parts per million of CO_2, an accelerated shift away from fossil fuels is needed.[89]

Substantially reducing carbon emissions will mean an end to the fossil fuel energy economy as we currently know it.

Until recently, most official energy projections denied that such a transition was possible. Analysts were lulled by the fact that basic energy systems had changed little in recent decades—retaining a basic reliance on steam cycle power plants, internal combustion engines, and liquid petroleum fuels. But this vision of a static energy system harkens back to the views of U.S. Commissioner of Patents Charles Duell, who argued in 1899 that his office should be closed since "everything that can be invented has been invented." Such pessimism is out of step with today's world of microchips and nanoseconds.[90]

Recent studies have presented converging sketches of a potentially new and more sustainable energy path the world could follow during the next half century. They envision a highly efficient, decentralized energy system relying on a new generation of high-tech, mass-produced conversion systems that turn abundant supplies of solar energy, wind power, biomass, and a limited amount of fossil fuels into useful forms of energy such as electricity and hydrogen.[91]

Studies by national governments, international agencies, and private companies concur that energy efficiency levels can be greatly improved. Using already identified technologies, the efficiency of automobiles and appliances can be doubled or tripled, household and commercial building energy use can be reduced by two-thirds or more, and electricity generators can go from current efficiencies of 30-35 percent to 50-60 percent. Small-scale power systems located inside buildings can turn 90 percent of the fuel they consume into electricity and usable heat. In developing countries, where much of the industrial equipment in use today is outdated and inefficient, efficiency levels can be increased even more dramatically.[92]

Extensive reliance on intermittent renewable energy sources will not be achieved as easily. Rather, it will require a burst of innovation and accelerated introduction of new technologies—analogous to the technological revolutions that ushered in electricity and automobiles between 1890 and 1910, or to the electronic revolution that followed the Second World War.[93]

Already, however, solar and wind energy technologies appear to be entering a "takeoff" phase of the kind that personal computers experienced in the early 1980s. The annual production of photovoltaic cells that turn sunlight directly into electricity has expanded nearly fourfold in the past decade, while their cost has fallen by half. Annual installation of wind turbines increased sixfold between 1990 and 1995, and costs fell so far that wind power is now as inexpensive as coal-based electricity in some regions.[94]

As with other mass-produced technologies, market growth and cost reduction go hand in hand. As sales expand, manufacturers invest in new technologies, gear up high-volume assembly lines, and lower their prices, which in turn allows them to expand their markets further. By following this path, solar and wind energy technologies that today provide less than 1 percent of the world's electricity could begin to displace large amounts of fossil fuels in a decade or two, according to several recent studies. Resource availability will not be a problem. The 2.5 million exajoules of solar energy that reach the earth's surface each year is

6,000 times as much energy as is used by human societies worldwide. Some of this solar energy is then converted to wind and biomass energy. Cost, not resource size, is therefore the key to making solar and wind power the foundations of a new energy system.[95]

Increased reliance on natural gas is another key element of strategies to stabilize the climate, according to recent studies. Gas is not only the least carbon-intensive of the fossil fuels, it also lends itself to efficient applications that can further reduce carbon emissions, including new refrigeration technologies, small-scale power generators using modified jet turbines and diesel engines, and a new device called a *fuel cell* that efficiently converts natural gas into electricity without burning it. One calculation shows that simply replacing all U.S. coal-fired power plants with gas-fired cogenerators could reduce national carbon emissions by 25 percent.[96]

Estimates of the earth's natural gas resources have expanded in recent years, suggesting that global production—thought in the 1980s to be near its all-time peak—will almost certainly double, and could more than triple by the middle of the next century. In most countries, reliance on natural gas is already rising far more rapidly than use of other fossil fuels. During the next few decades, gas could displace most current uses of coal as well as some oil. Because of the physical and chemical similarities between natural gas and hydroden, natural gas can form a sort of bridge to hydrogen-based energy systems. Hydrogen in turn can be derived from water, using widely available energy sources such as wind and solar power.[97]

Even industry groups have begun to consider the feasibility of less fossil-fuel-dependent energy paths. A study by the strategic planning department of Royal Dutch Shell in 1994 concluded that continued energy reforms and the liberalization of energy markets, together with stronger environmental laws, could unleash powerful new technologies and allow us to rely mainly on renewable energy sources by the end of the 21st century.[98]

The World Energy Council, which represents leading energy companies, published a study in 1995 that reflects a new and more hopeful perspective for this fossil-fuel-dominated organization. Its new "ecologically driven" scenario would hold total energy use to 60 percent over the current level by 2050, compared to the 200-300 percent increase projected by the group's standard scenarios. Renewable resources would claim 40 percent of the market—more than oil, coal, and nuclear power combined. Carbon dioxide emissions, meanwhile, would decline 10-15 percent by 2050 and 50-70 percent by 2100, stabilizing concentrations of carbon dioxide at between 450 and 500 parts per million—within reach of the goals described earlier.[99]

By following such a path, an efficient, decentralized energy system relying heavily on renewables could emerge relatively seamlessly out of today's system, using existing and expanded gas pipeline networks to transport solar- and wind-generated hydrogen, and employing electric power grids to transmit electricity produced from remote windfarms as well as rooftop-mounted solar cells.[100]

The debate is now shifting from whether a low-carbon energy system is feasible to whether it is affordable—a question that still provokes controversy. On one side are many economists and fossil fuel lobbyists who use economic models to suggest that holding carbon emissions steady during the next few decades could cost hundreds of billions of dollars and cut economic output by 1.5-2.5 percent. These conclusions are challenged by other economists and many natural and social scientists who view the studies as flawed. Most of the economic models used by the pessimists assume that the current path is unfettered by market barriers, and that technology is relatively static. Moreover, they assume that the shift in energy paths would be forced by a steep carbon tax—not because that is the most efficacious approach but because it is the easiest policy to model. They neglect the potential to remove barriers to the introduction of new energy technologies, which could reduce emissions at much lower cost.[101]

A.J. McMichael, a professor at the London School of Hygiene and Tropical Medicine, wrote in *Science* that "attempting to reduce the complexities of climate change, its impact on the biosphere, and the implications for human experiences to market-based values, and then to pass the bill to future generations in the hope that they will appreciate our far-sighted capital accumulation strategy, is not a solution." The IPCC's 1995 report echoed this conclusion, noting that there is now strong economic justification for shifting to a range of new investments in sustainable energy technologies and other means of reducing greenhouse gas emissions.[102]

Early efforts to estimate the cost of reducing pollution almost always turn out to be too high. It was once projected, for example, that reducing sulfur emissions in the United States would cost nearly $700 per ton, whereas in early 1996, the permits to emit sulfur that are auctioned by the U.S. Environmental Protection Agency were trading at less than $70 per ton. And in replacing ozone-depleting chemicals, refrigerator manufacturers had to devise new technologies that turned out to be more energy efficient and therefore economical to operate than the ones they replaced.[103]

Early efforts to estimate the cost of reducing pollution almost always turn out to be too high.

Just as it was impossible to prove in 1980 that powerful *and* affordable personal computers could be developed, so is it difficult today to know for sure what a new energy economy might cost. Still, the recent pace of progress in a number of key technologies, as well as the lower operating costs of some of the new systems provide reason for optimism. Today, for example, combined-cycle power plants that have both steam turbines and gas turbines—which were not even on the list of options for most utility executives a decade ago—cost less than half as much to build as a standard coal plant, while emitting one-third as much carbon dioxide.[104]

Once market barriers are removed, new energy technologies are commercialized, and environmental costs recognized, a low-carbon economy may turn out to be more cost-effective than today's high-carbon one. Of course, building a new energy system will require sizable up-front investments. But over time, the costs can be offset by the lower operating and fuel costs of the new systems, as well as associated environmental benefits such as lower health care bills that will result from a less-polluting energy system.[105]

One area of continuing controversy surrounds the optimal timing of the shift ahead. Three leading scientists suggested in a January 1996 article in *Nature* that it might be more economical to allow emissions to rise somewhat in the next two to three decades and then reduce them more rapidly in later decades as more cost-effective new technologies become available. (The two emission paths by which the 450 ppm and 550 ppm targets can be achieved are shown in Figure 7.) The study notes that because it is cumulative emissions that most influence climate change, lower emissions in the more distant future can offset near-term emissions. Although the authors went on to say that their analysis should *not* be interpreted as suggesting a "do-nothing" or "wait-and-see" policy, it has provided ammunition for those who argue just that.[106]

It is correct, indeed obvious, that there are several different emission trajectories that can get the world to a low-carbon energy system; but the more important question is how to minimize the risks we now face. Since scientists do not yet know exactly what level of greenhouse gas concentrations we and future generations can live with, there is a real danger of overshooting critical limits. Delay could be dangerous since it will inevitably take time to prepare the groundwork for the kind of rapid change in energy and transport systems that may soon be needed. And as Figure 7 indicates, the longer action is delayed, the sharper and more disruptive the eventual shift in emission trends will have to be.[107]

One critical priority is accelerating the commercial development of new, low-carbon energy technologies since it takes time for new systems to displace a significant portion of the devices currently in use. To achieve any reasonably safe emissions trajectory, the market for fuel cells, photovoltaics, electric vehicles, and other critical technologies will need to be accelerated soon. The highest priorities are the facilities that last the longest: a large office building or coal-fired power plant built today could be contributing to global warming a half century from now. During the next three decades, most of today's power plants, refineries, and factories will be replaced, while automobiles and some appliances will turn over two to three times. These turnovers are opportunities to lower greenhouse gas emissions that should not be missed.

Unlike most problems we face, greenhouse warming is a phenomenon that will take centuries to reverse—because of the long time lags involved. And if we suddenly discover a decade from now that the limits need to be lowered, a costly crash effort may be needed—replacing existing power plants and other equipment long before they become obsolete, for example. Therefore, the precautionary principle suggests that we should err on the side of conservatism and begin the inevitable move away from fossil fuels now.

Rio, Berlin, and Beyond

Global climate change will not be slowed with a simple law or regulation. More than any other environmental problem, climate change is woven into the very structure of today's societies. Each factory, office building, and automobile contributes to greenhouse warming. Major changes in technology, infrastructure, and even life-style are needed to slow it. Such changes will depend on major policy reforms that not only reduce dependence on fossil fuels, but also alter the way cities are designed, agriculture is practiced, and forest lands are managed.

One of the greatest challenges is the fact that climate policy cuts across so many lines of governmental jurisdiction. While it is one of the few issues that is truly global in scope—a kilogram of carbon dioxide emitted in Canada has exactly the same consequence as one released in Sri Lanka—essential policy changes need to occur at the national, state, and local levels. Energy taxes, for example, are generally levied nationally, electric utilities are often regulated by states, and road and building codes are usually set by cities and towns. Effectively slowing the pace of climate change will therefore require cooperation and coordination among all levels of government, with broad guidelines and goals set at the international level, and much of the policy implementation occurring closer to where people live and work.

The first Conference of the Parties to the climate convention, held in Berlin in April 1995, was a turning point in negotiations. After a nearly three-year hiatus following the treaty's adoption, delegates in Berlin finally went beyond wrangling over procedural matters and faced the large gap between the treaty's lofty goals and the stark reality of growing greenhouse gas emissions. After two weeks of hard bargaining, the Conference of the Parties adopted the "Berlin Mandate," which charges treaty members with adopting a protocol to the convention aimed for the first time at reducing these emissions. In the awkward language of this mandate, the protocol is to "set quantified limitation and reduction objectives [for industrial countries] within specified time frames, such as 2005, 2010, and 2020."[108]

Delegates also agreed in Berlin to consider a range of specific measures to reduce emissions, and to allow some countries to launch a series of pilot projects to transfer less carbon-intensive technologies between nations. In addition, the first Conference of the Parties established guidelines for compiling, submitting, and reviewing national reports on greenhouse gas inventories, projections, and policies, but failed—due largely to the opposition of oil exporting countries—to adopt rules of procedure under which the convention itself will operate.[109]

The Berlin agreements may not ensure that the climate is stabilized, but given the controversy and contention that have marked the treaty process to date, and the sense of foreboding that preceded the Berlin meeting, the renewed commitment is encouraging. Together with the strengthened IPCC scientific assessment issued later in the year, Berlin injected new energy into the international climate policy process. U.K. Environment Secretary John Gummer summed up the achievement: "The deal means we have a real chance of avoiding the worst effects of climate change."[110]

Still, the Berlin Mandate was a delicate compromise that papers over remaining differences about how stringent the treaty's targets should be. At meetings in Geneva in early 1996, German negotiators continued to push for tough, legally binding limits for industrial country CO_2 emissions—a 10 percent reduction by 2005 and a 15-20 percent cutback by 2010—as well as unspecified targets for other greenhouse gases. Other industrial countries, including the United States, remain reluctant to agree to such strict limitations.[111]

Unfortunately this new debate over binding targets and timetables is becoming bogged down in the kind of contentious deliberations that have plagued such discussions since 1990. Experience with the treaty's existing targets is not encouraging. Some nations have creatively interpreted their obligations—shifting to a new base year, focusing on emissions per person rather than total emissions, and juggling the relative contribution of various gases. Meanwhile, carbon dioxide emissions in some countries have already overshot the treaty's target, while others have fallen sharply—mainly as a result of broader economic shifts rather than national climate policies.[112]

Additional emissions limits are essential to the long-run effectiveness of the climate convention, but the time may not yet be ripe for restrictions that are sufficiently ambitious, specific, *and* legally binding. Researchers at the

International Institute for Applied Systems Analysis have suggested as an interim step that stronger voluntary targets and timetables be adopted in 1997, following the model that was successfully used in the 1980s to gradually strengthen agreements on reducing transboundary air pollution and North Sea pollution. They argue that these voluntary targets, which often involve informal, self-selected "clubs" of countries, can produce much more demanding goals and lead to immediate environmental gains while building the basis for subsequent targets that are legally binding.[113]

This interim approach could also strengthen the convention's currently weak reporting and review mechanisms. It would force governments to review the effectiveness of their climate plans and hold them up to domestic political as well as international scrutiny, which might spur them to act. Ideally, the new targets should be differentiated according to current emission rates, and should cover developing countries too—aimed at reducing the rate at which their emissions grow.

Two broad global approaches to limiting greenhouse gas emissions have been proposed. The first is a system of carbon dioxide taxes that would be used to discourage combustion of fossil fuels. It would function as a sort of energy tax that increases the price of gasoline, electricity, and other forms of energy to the consumer. Economists argue that such a tax would follow the "polluter pays" principle, requiring consumers to pay the full environmental cost of the fuels they use.

In order to slow fossil fuel use significantly, a sizable tax on carbon dioxide would be required. A levy of $200 per ton of carbon on all fossil fuels—equivalent to $0.14 per liter of gasoline ($0.54 per gallon)—would raise $1,200 billion annually, equivalent to roughly 16 percent of current global tax receipts. If such a tax were added suddenly and without compensating offsets, it would slow the economy, but most economists argue that the taxes should be raised gradually, matched by a decline in other taxes—on income, labor, or capital, for example. Such a tax shift would tend to accelerate the development of new energy technologies and also spur sales and job creation in industries like telecommunications and insurance. Their growth could offset losses in industries such as chemicals and steel. The result might well be a surge in investment that could boost the economy as a whole.[114]

Denmark, Finland, the Netherlands, Norway, and Sweden have adopted small carbon dioxide taxes already, but the idea has met considerable resistance elsewhere, chiefly by industries that feel it would put them at a disadvantage. Their main objection is that a carbon dioxide tax adopted by one country would help their foreign competitors. In response to this concern, the European Union has considered a union-wide energy or carbon dioxide tax, but the proposal seems unlikely to be adopted anytime soon.[115]

Denmark, Finland, the Netherlands, Norway, and Sweden have adopted small carbon dioxide taxes already.

Another approach to limiting greenhouse gas emissions is to impose binding limits and allow trading of emission permits among companies as well as nations. A similar mechanism was included in the acid rain section of the U.S. Clean Air Act amendments of 1990. Companies that emit large amounts of sulfur dioxide are required either to reduce their emissions below a specified cap, or else to purchase emissions permits from other companies that go below their legal limits. The trading system has reduced the cost of lowering sulfur emissions. Its effectiveness is shown by the fact that, over time, the price of the tradeable permits has steadily fallen.[116]

The price of emissions permits in a CO_2 trading program would be determined by the market, which would be influenced by the amount of emissions reductions called for and by how much it would cost industry to meet those requirements. Facilities that continued to emit carbon above their allowed cap would have to purchase permits—

which would have an effect similar to that of a tax. However, some analysts argue that this approach is more likely to be politically feasible than a tax, and would do a better job of catching the eye of plant managers and other executives who are in a position to reduce emissions. One problem with most such schemes is that the emissions permits would apply only to major polluters, doing little to encourage millions of small emitters—owners of farms, homes, and automobiles, for example—to reduce their emissions.[117]

As a first step in emissions trading, some industrial countries have proposed a concept known as "joint implementation" that would allow countries to offset their own emissions by investing in clean energy projects or reforestation elsewhere. The idea has strong support among companies in industrial nations that release large quantities of carbon dioxide and want to develop early offsets against the possibility that they will be required at some point in the future to limit their emissions. Advocates such as the U.S. and Norwegian governments argue that the most cost-effective emissions reductions can be achieved in developing nations, whose energy systems are generally inefficient, and whose emissions are growing rapidly. They support joint implementation as a way to promote technology exports and take the pressure off their own carbon-intensive sectors.[118]

The U.S., Dutch, and Norwegian governments have begun encouraging pilot-phase joint implementation projects carried out by their own domestic companies. The projects signed up under the U.S. program so far—which are all privately financed and are not accruing formal credits—include a geothermal energy project in Nicaragua, a small wind farm and reforestation project in Costa Rica, and an effort to reduce methane leaks from a pipeline in Russia. Central American and East European countries have seized on joint implementation as a way of financing important energy and forestry projects. So far, it is too early to know how successful any of them will be.[119]

Elsewhere, most developing nations oppose joint implementation, arguing that it will allow richer nations to avoid responsibility for reducing their own emissions, which represent a far larger part of the climate problem than those of poor countries. Charges of "environmental colonialism" have come close to turning climate negotiations into another skirmish in the North-South tug-of-war that marks many U.N. processes. These equity arguments have been joined by practical questions. U.S. carbon emissions are projected to rise by 140 million tons between 1990 and 2000—equivalent to the current annual emissions of Mexico and Brazil combined. Offsetting those emissions would require planting trees over an area the size of Italy at a cost of many billions of dollars. Such sums might well be better spent improving the still lackluster energy efficiency of many U.S. buildings and factories.[120]

These calculations suggest that the climate problem will not be solved just by encouraging industrial countries to offset their emissions via tree-planting programs in developing countries. In fact, recent studies question whether large-scale planting efforts would result in sequestering nearly as much carbon as academic assessments promise. However, efforts to improve the energy efficiency of developing countries could be a useful supplement to emission reduction efforts in industrial countries. Since developing nations would likely be allowed to increase their emissions under any system of global limits, industrial countries would need to purchase some of those allowances, providing financial infusions to the energy, transportation, agriculture, and forestry sectors of developing nations. As an interim compromise, negotiators in Berlin decided to permit pilot-phase joint implementation projects through 1999, but the investing countries will not be permitted to accumulate credits for the reductions achieved during this period.[121]

The Netherlands, Norway, Sweden, and the United States continue nevertheless to launch joint implementation projects, and interest in emission trading schemes is still growing. Some proponents have gone beyond assembling loose collections of energy and forestry projects to propose strategies to allow emissions trading on a scale large enough to make a dent in the overall problem. One plan would start by setting up detailed standards for certifying greenhouse gas offsets and authorizing auditing firms to certify the results. Participating companies would have to submit annual audits of their firms' emissions. Under this proposal, emissions would first be permitted bilaterally among countries at similar levels of economic development. Possibilities include Norway-Sweden, Canada-United States, and Australia-New Zealand. The list could later be expanded to include broader collections of countries, eventually extending between industrial and developing nations.[122]

Although such schemes would not reduce greenhouse gas emissions greatly in the near future, their impact would likely grow over time. Drawing corporations into the process of tracking their emissions and looking for low-cost reduction options could well spur hundreds of important initiatives. Moreover, such programs might lead eventually to binding emissions limits of the sort that have underpinned all successful emissions trading systems so far.

Even as these approaches are being developed, the climate convention can be used to encourage narrower but still important policy reforms that will lower greenhouse gas emissions, particularly from major contributors such as automobiles, home appliances, power plants, and landfills. This sector-by-sector approach is sometimes criticized as inadequate to the scale of the climate problem, but it may be the most effective means of lowering the market barriers to new technologies—some of which would not be much affected by a carbon tax or emissions trading system. Some argue that such measures should be left entirely to national governments to decide, but the treaty process could help standardize such norms internationally, reducing the cost to industry of compliance, particularly for those products that are heavily traded.

Among the "policies and measures" that could be encouraged by the convention are energy efficiency standards. Many industrial countries have established efficiency requirements for buildings, appliances, and automobiles—greatly reducing energy use. However, many developing countries do not yet have such standards, and so are unnecessarily wasting energy and money. The climate convention could adopt energy efficiency standards—either legally binding ones or simply model codes—for a range of important technologies. Taking a similar approach, the German government has suggested that the European Union adopt a new car fuel economy requirement of 5 liters/100 km (~45 mpg).[123]

The climate convention could also help to certify and improve the range of voluntary industry programs that have proliferated in recent years. The Netherlands has demonstrated that major changes in industry practices can be achieved via tough voluntary environmental covenants negotiated with government officials. But the U.S. government and others have allowed companies to sign up to minimal voluntary programs that are unlikely to have much effect on emissions. Such programs would benefit from standardized guidelines. The International Standards Organization (ISO) is currently developing a new set of standards for industrial environmental practices called ISO 14000 that could be a useful step in this direction.[124]

Major changes in industry practices can be achieved via tough voluntary environmental covenants.

The climate convention could also be used to establish market-incentive programs that would spur the adoption of new technologies. Tax incentives in India, generous power purchase prices in Germany, and "golden carrots" for the bulk purchase of innovative energy devices in the United States are among the creative programs devised to accelerate the commercial development and use of important new energy technologies. Analysis and certification of such programs—distinguishing them from token efforts that have yielded little—would help countries that wish to promote new approaches to energy and transport policy but are uncertain how to do so.[125]

Financial assistance for developing countries is another key element of the climate convention, but one that has so far received less support than needed. Treaty members have agreed that the Global Environment Facility (GEF), set up under the joint management of the World Bank, the United Nations Development Fund, and the United Nations Environment Programme, will be the chief funding mechanism for the treaty. By late 1995, the GEF had allocated $328 million to projects that are dedicated to slowing climate change. Much of this funding has gone to the preparation of national greenhouse gas inventories and climate plans. Other GEF climate projects include a power plant using sugar cane wastes in Brazil, a factory to produce more efficient industrial boilers in China, an energy efficiency fund in Hungary, and a solar thermal power plant in India.[126]

It is too early to judge most of these projects, but even if they are successful, the GEF's resources are miniscule compared to the tens of billions of dollars invested in exploitation of fossil fuels in developing countries each year. At its current scale, the GEF fund is mainly useful as a means of demonstrating new technologies and strategies. The real money is at the World Bank and other multilateral funding organizations, which still devote less than 5 percent of their energy lending to energy efficiency projects. Efforts to reform these priorities continue to meet resistance from client governments and World Bank staff. The World Bank has recently begun to require assessments of the climate implications of its energy and transport projects, but still has no formal guidelines for determining whether a given project is acceptable. Clearer climate standards are a high priority for all multilateral banks.[127]

None of these policies are likely to be adopted unless heavy political resistance is overcome. Fossil-fuel-dependent industries, oil-exporting nations, and hired-gun scientists have confused the public and thrown up a host of procedural barriers. These interests, particularly those based in the United States, have spent millions of dollars lobbying against tightening the climate convention in recent years; their position papers have found their way into official Saudi and Kuwaiti statements and, in one embarrassing episode, a speech by a U.S. government official. Although most of the carbon lobby's machinations are transparently obstructionist, they have slowed a process that normally operates by consensus and is easily bogged down by even one hold-out nation. It is Kuwait and Saudi Arabia, for example, that have prevented treaty members from even adopting rules of procedure.[128]

Still, beginning in Berlin in 1995, a more progressive confluence of political forces began to assert itself—prominently featuring the insurance and banking industries. As a business on the front lines of society's most risky activities, the insurance industry has a long tradition of spurring policy changes to help reduce society's risks. In the United States, for example, the industry's experience with fire-related claims led it to lobby for stricter building codes that reduce the frequency of fires. Similarly, insurers have fought since the early 1970s for tougher safety standards for automobiles—often battling directly with auto industry lobbyists. The resulting regulations requiring crash-resistant bumpers, seat belts, and air bags have saved tens of thousands of lives—and prevented billions of dollars in insurance losses.[129]

With this history in mind, a growing number of industry leaders now argue that insurance companies should take a more direct role in global climate policy. In a follow-up report on the Berlin conference, a representative of Lloyd's of London predicted "that the insurance industry would have to take some initiatives by itself, or along with the banking industry."[130]

Such efforts are being complemented by Sustainable Energy Business Councils that were formed in the United States, Australia, and Europe during the last few years. Companies that sell everything from home appliances to

TABLE 3

International Climate Alliances, Berlin, 1995

Name	Sample Membership	Concerns	Positions
Carbon Club	U.S., Canada, Australia, New Zealand, Japan	Cost of breaking away from heavy reliance on fossil-fuel-based economy; worry that reducing emissions will slow the economy, creating unemployment.	Opposed to quantified reductions below 1990 levels; prefer a cumulative emission budget approach; opposed to harmonization of policies and measures; support joint implementation projects.
Euro-Greens	Germany, Netherlands, Denmark, U.K.	Economic and social costs of climate change, including floods, sea level rise, and environmental refugees.	Favor protocol with specific commitments to emission "targets and timetables" going beyond 2000; proposed new car efficiency standards, energy taxes, mandatory harmonized policies.

OPEC	Saudi Arabia, Kuwait, Venezuela	Threat to oil revenues.	Opposed to quantified reduction targets, harmonized policies, and any substantive, or even procedural action on climate change.
AOSIS	Maldives, Kiribati, Mauritius	Damage from climate change, particularly sea level rise, coastal destruction, damage to economy.	Favor protocol that obligates industrial countries to reduce CO_2 emissions by 20% over 1990 levels by 2005.
Developing Country Green Group	Most developing countries; leaders include Philippines, India, China, Brazil	Impacts of climate change such as sea level rise, decreasing agricultural yields, spread of infectious diseases, collapse of fisheries; also concerned that emissions reductions could limit their capacity to develop economically.	Favor protocol that obligates industrial countries to reduce CO_2 emissions 20% from 1990 levels by 2005; opposed to any new commitments for developing countries; opposed to credits to industrial countries for mitigation measures undertaken in developing nations.

Source: Various reports and articles.

building insulation, wind power, and natural gas now believe that they could gain huge markets from the shift to new energy technologies, and are finding common cause in efforts to make the climate convention more effective. Their businesses could grow even faster if, as some have suggested, insurance companies would jettison a portion of their extensive holdings in oil and coal, and invest those funds in new energy technologies.

At the national level, a strong new North-South climate alliance began to emerge in Berlin. (See Table 3.) German environmental groups and the public at large put heavy pressure on the host government throughout the conference, and the European Union emerged as a champion of stronger climate policies. Their efforts were supported by scores of local officials from developing and industrial countries, representing 150 cities that are working to reduce their own emissions by focusing on transportation planning and more efficient buildings. At a press conference chaired by the Mayor of Toronto, these officials were critical of the lethargic efforts of national governments.[131]

Another voice for action that asserted itself in Berlin was the developing world, led, ironically, by its smallest members. For several years, the Alliance of Small Island States (AOSIS)—composed of 36 tiny nations—has been active in climate negotiations. In Berlin, AOSIS tabled a proposal during the first week that would commit industrial countries to reduce their emissions 20 percent.[132]

That proposal drew unexpected support in Berlin from most of the Group of 77, an alliance of over 100 developing countries that often votes as a bloc. The G-77 was chaired in Berlin by Philippines Senator Angel Alvarez, who had recently participated in a study that demonstrated the enormous economic damage that climate change might cause in Asia. The Philippines was soon joined by Malaysia, Bangladesh, and other concerned countries that joined Alvarez' call for sharp reductions in industrial country greenhouse gas emissions.[133]

By the end of the conference's first week, this "Green

Group" was joined by most of the G-77, including China, India, and Brazil—countries that once viewed climate change as a "rich countries' problem." Only the oil-exporting countries held out. The Green Group joined forces with Germany and other European countries, which were already battling with the United States and other members of the "carbon club" over the toughness of the Berlin Mandate. Outnumbered, the U.S. government backed down on many key points in an all-night bargaining session on the last day of the conference.[134]

These unexpected developments in Berlin provide cause for hope. Policies that may seem beyond reach today can be adopted by acclamation tomorrow as the political winds shift. North-South partnerships are key to such progress. Industrial countries have a moral and practical responsibility to lead. After all, they largely created the climate problem, and they have the technology and money needed to solve it. But unless developing nations change course well before adopting a full-fledged fossil fuel economy, the climate cannot be stabilized. The two groups have a common interest in developing cost-effective energy alternatives as soon as possible.

Efforts to head off ozone depletion were also moving at a slow pace in the early 1980s—until dramatic new evidence of a "hole" in the Antarctic ozone layer emerged at about the same time that leading chemical companies realized that they could make even more money selling substitutes for older chemicals. In just a short period, these companies ended their opposition to phasing out the older chemicals. Developing countries, meanwhile, were offered a special ozone fund to assist their transition to substitutes. Soon, world leaders were calling for swift action, and in 1990, the treaty was revised to call for an end to production of the most damaging chemicals by 2000. Since then, output of those chemicals has fallen 70 percent, and scientists agree that the threat posed by them is on the way to being eliminated.[135]

We are still a long way from stabilizing the global climate, a far more complex challenge than repairing the ozone layer. Even with quick action, some greenhouse gases will linger in the atmosphere for centuries. Still, close observers note that a climate of hope has crept into the negotiations recently. Insurance companies, small island nations, and others with major interests in a stable climate have re-shaped the diplomatic playing field. Finally, the time for serious policymaking may be at hand.

Notes

1. J.T. Houghton et al., eds., *Climate Change 1995: The Science of Climate Change, Contribution of Working Group I to the Second Assessment Report of the Intergovernmental Panel on Climate Change* (New York: Cambridge University Press, 1996); Tom M.L. Wigley, "A Successful Prediction?" *Nature,* August 10, 1995.

2. These specific trends and the references to support them are described later in this paper. The same holds true of most issues discussed in the introduction, all of which are covered in more detail later.

3. Franklin W. Nutter, speech at Conference on Financing Strategies for Renewable Energy and Efficiency, New York, N.Y., May 11, 1994.

4. J.T. Houghton, B.A. Callender, and S.K. Varney, eds., *Climate Change 1992: The Supplementary Report to the IPCC Scientific Assessment* (New York: Cambridge University Press, 1992); Houghton et al., eds., op. cit. note 1.

5. Svante Arrhenius, "On the Influence of Carbonic Acid in the Air Upon the Temperature of the Ground," *Phil. Magazine,* 1896.

6. C.D. Keeling and T.P. Whorf, "Atmospheric CO_2 Records from Sites in the SIO Air Sampling Network," in Thomas A. Boden et al., eds., *Trends '93: A Compendium of Data on Global Change* (Oak Ridge, Tenn.: Oak Ridge National Laboratory, 1994); Timothy Whorf, Scripps Institution of Oceanography, La Jolla, CA, private communication and printout, February 5, 1996; Thomas E. Graedel and Paul J. Crutzen, *Atmosphere, Climate, and Change* (New York: Scientific American Library, 1995); C.D. Keeling et al., "Interannual Extremes in the Rate of Rise of Atmospheric Carbon Dioxide Since 1980", *Nature,* 22 June 1995.

7. T. Blunier, "Historical CH_4 Record from the Eurocore Ice Core at Summit, Greenland" in Boden et al., eds., op. cit. note 6; Houghton, et al., eds., op. cit. note 1; Graedel and Crutzen, op. cit. note 6; V. Ramanathan et al., "Trace Gas Trends and Their Potential Role in Climate Change," *Journal of Geophysical Research,* June 20, 1985; heat trapping gas estimate is based on radiative forcing figures for various greenhouse gases contained in IPCC, *Climate Change: The IPCC Scientific Assessment* (New York: Cambridge University Press, 1990).

8. Text and Figure 1 are from H. Wilson and J. Hansen, "Global and Hemispheric Temperature Anomalies from Instrumental Surface Air Temperature Records," in Boden et al., eds., op. cit. note 6; James Hansen et al., "Table of Global-Mean Monthly, Annual and Seasonal Land-Ocean Temperature Index, 1950-Present," Goddard Institute for Space Studies Surface Air Temperature Analyses, as posted at http://www.giss.nasa.gov/Data/GISTEMP, May, 1996; Phil Jones et al., "University of East Anglia Land Air Temperatures With Sea Surface

Temperatures," Climatic Research Unit, as posted at http://www.cru.uea.ac.ukc/cru/press/pj9601/data.htm, January 1996.

9. "Stormy Weather Ahead," *Economist*, March 23, 1996; Boyce Rensberger, "Climatology: What's Hot, What's Not," *Washington Post*, January 8, 1996; "Top 10 Global Warming Myths," *World Climate Report* (edited by Patrick Michaels), Vol. 1, No. 5, 1995; Robert C. Balling, Jr., "Keep Cool About Global Warming," *Wall Street Journal*, October 16, 1995; James Hansen et al., "Satellite and Surface Temperature Data At Odds?" *Climatic Change*, Vol. 30, 1995; John R. Christy and Richard T. McNider, "Satellite Greenhouse Signal," *Nature*, January 27, 1994; Michael MacCracken, "The Evidence Mounts Up," *Nature*, Vol. 376, August 24, 1995; William K. Stevens, "'95 The Hottest Year On Record As the Global Trend Keeps Up," *New York Times*, January 4, 1996; Malcolm W. Browne, "Most Precise Gauge Yet Points to Global Warming," *New York Times*, December 12, 1994.

10. John Firor, *The Changing Atmosphere: A Global Challenge* (New Haven: Yale University Press, 1990).

11. The Hadley Centre for Climate Prediction and Research *Modeling Climate Change: 1860-2050* (Bracknell, UK: The Meteorological Office, February 1995); Robert Matthews, "The Rise and Rise of Global Warming," *New Scientist*, November 26, 1994; Tom M.L. Wigley, "A Successful Prediction," *Nature*, August 10, 1995; J.F.B. Mitchell et al., "Climate Response to Increasing Levels of Greenhouse Gases and Sulphate Aerosols," ibid.; M. Patrick McCormick, Larry W. Thomason, and Charles R. Trepte, "Atmospheric Effects of the Mt. Pinatubo Eruption," *Nature*, February 2, 1995; Figure 2 from Mitchell et al., op. cit. this note and from data supplied by Bob Davis, Hadley Centre for Climate Prediction and Research, Bracknell, U.K., October 5, 1995.

12. Houghton et al., eds., op. cit. note 1; Stephen H. Schneider, "Detecting Climatic Change Signals: Are There Any 'Fingerprints'?" *Science*, January 21, 1994; Richard Monastery, "Dusting the Climate for Fingerprints," *Science News*, June 10, 1995; D.G. Vaughan and C.S.M. Doake, "Recent Atmospheric Warming and Retreat of Ice Shelves On the Antarctic Peninsula," *Nature*, January 25, 1996; J.C. King, "Recent Climate Variability in the Vicinity of the Antarctic Peninsula," *International Journal of Climatology*, May 1994.

13. Keith R. Briffa et al., "Unusual Twentieth-Century Summer Warmth in a 1,000-Year Temperature Record from Siberia," *Nature*, July 13, 1995; Molly Moore, "New Delhi Cools Off—At 98 Degrees," *Washington Post*, June 14, 1994; Steve Newman, "Earth Week," *San Francisco Chronicle*, January 13, 1996; David Roberts, "The Iceman: Lone Voyager from the Copper Age," *National Geographic*, June 1993.

14. William K. Stevens, "Scientists Say Earth's Warming Could Set Off Wide

Disruptions," *New York Times,* September 18, 1995; Northern Finland information from Charles Petit, "New Hints of Global Warming," *San Francisco Chronicle,* April 17, 1995; Ross Gelbspan, "The Heat is On," *Harper's Magazine,* December 1995.

15. Thomas R. Karl, Richard W. Knight, and Neil Plummer, "Trends in High Frequency Climate Variability In the Twentieth Century," *Nature,* September 21, 1995.

16. David J. Thomson, "The Seasons, Global Temperature, and Precession," *Science,* April 7, 1995; "Springtime for Scientists in Georgia," *Economist,* February 25, 1995.

17. Houghton et al., eds., op. cit. note 1; Karl quoted in "Reading the Patterns," *Economist,* April 1, 1995; Thomas R. Karl et al., "Trends in U.S. Climate During the Twentieth Century," *Consequences,* Spring 1995; Karl et al., op. cit. note 15; Hasselmann cited in Monastery, op. cit. note 12.

18. Balling, op. cit. note 9; Richard A. Kerr, "Greenhouse Skeptic Out in the Cold," *Science,* December 1, 1989; idem, "Greenhouse Science Survives Skeptics," *Science,* May 22, 1992; Patrick Michaels, *Sound and Fury: Science and Politics of Global Warming* (Washington, D.C.: Cato Institute, 1992); Richard Lindzen, "Absence of Scientific Basis," *Research and Exploration,* Spring 1993. For funding by the fossil fuel industry, see masthead of *World Climate Report* (edited by Patrick Michaels); Gelbspan, op. cit. note 14; Ozone Action, *Ties That Blind: Case Studies of Corporate Influence on Climate Change Policy* (Washington, D.C., March 1996); Gary Lee, "Industry Funds Global-Warming Skeptics," *Washington Post,* March 21, 1996.

19. Houghton et al., eds., op. cit. note 1.

20. Houghton et al., eds., op. cit. note 1; Figure 3 from J. Jouzel et al., "Vostok Isotopic Temperature Record," in Boden et al., eds., op. cit. note 6.

21. Stephen H. Schneider, *Global Warming: Are We Entering the Greenhouse Century?* (San Francisco: Sierra Club Books, 1989).

22. George M. Woodwell and Fred T. Mackenzie, eds., *Biotic Feedbacks in the Global Climatic System: Will the Warming Feed the Warming?* (New York: Oxford University Press, 1995); Pieter P. Tans and Peter S. Bakwin, "Climate Change and Carbon Dioxide," *Ambio,* September 1995; Philip Newton, "Perspectives Past and Present," *Nature,* November 2, 1995; Deborah MacKenzie, "Where Has All the Carbon Gone?" *New Scientist,* January 8, 1994.

23. Houghton et al., eds., op. cit. note 1; A. Scott Denning, Inez Y. Fung, and David Randall, "Latitudinal Gradient of Atmospheric CO2 Exchange with Land Biota," *Nature,* July 20, 1995; P. Ciais et al., "A Large Northern Hemisphere Terrestrial CO2 Sink Indicated by the 13C/12C Ratio of

Atmospheric CO2," *Science,* August 25, 1995.

24. Raja S. Ganeshram et al., "Large Changes in Oceanic Nutrient Inventories from Glacial to Interglacial Periods," *Nature,* August 31, 1995; Louis A. Codispoti, "Is the Ocean Losing Nitrate?" *Nature,* August 31, 1995; Robert U. Ayres, William H. Schlesinger, and Robert H. Socolow, "Human Impacts on the Carbon and Nitrogen Cycles," in R. Socolow et al., eds., *Industrial Ecology and Global Change* (Cambridge: Cambridge University Press, 1994).

25. Houghton et al., eds., op. cit. note 1; R.T. Watson, M.C. Zinyowera, and R.H. Moss, eds., *Climate Change 1995: Impacts, Adaptation, and Mitigation of Climate Change—Scientific-Technical Analysis, Contribution of Working Group II to the Second Assessment Report of the Intergovernmental Panel on Climate Change* (New York: Cambridge University Press, 1996).

26. Sandra L. Postel, Gretchen C. Daily, and Paul R. Ehrlich, "Human Appropriation of Renewable Fresh Water," *Science,* February 9, 1996; World Bank, "Earth Faces Water Crisis," press release, Washington, D.C., August 6, 1995.

27. Watson et al., eds., op. cit. note 25; Lester R. Brown, "Facing Food Scarcity," *World Watch,* November/December, 1995.

28. Cynthia Rosenzweig and Martin L. Parry, "Potential Impact of Climate Change on World Food Supply," *Nature,* January 13, 1994; Woodwell and MacKenzie, eds., op. cit. note 22; William K. Stevens, "Scientists Say the Earth's Warming Could Set Off Wide Disruptions," *New York Times,* September 18, 1995; "Higher Temperatures, Carbon Dioxide Levels Could Harm Europe's Agriculture Production," *International Environment Reporter,* March 20, 1996.

29. Watson et al., eds., op. cit. note 25.

30. Ibid.

31. Watson et al., eds., op. cit. note 25; Robert L. Peters and Thomas E. Lovejoy, eds., *Global Warming and Biological Diversity* (New Haven: Yale University Press, 1992); Thomas J. Goreau and Raymond L. Hayes, "Coral Bleaching and Ocean 'Hot Spots'," *Ambio,* May 1994; Stephen C. Jameson, John W. McManus, and Mark D. Spalding, *State of the Reefs,* International Coral Reef Initiative Executive Secretariat Background Paper, May 1995.

32. Watson et al., eds., op. cit. note 25; Peters and Lovejoy, eds., op. cit. note 31.

33. Watson et al., eds., op. cit. note 25; Samuel Fankhauser, *Valuing Climate Change: The Economics of the Greenhouse* (London: Earthscan Publications, 1995); U.S. Climate Action Network, *Global Climate Change: U.S. Impacts* *and Solutions* (Washington, D.C., October 1995); William DiBenedetto, "Sun Stroke? Pacific Coast Salmon Face Menace of Global Warming," *Journal of Commerce*, January 3, 1996.

34. Watson et al., eds., op. cit. note 25.

35. Ibid.; Hamburg cited in Stevens, op. cit. note 14; Petit, op. cit. note 14.

36. Watson et al., eds., op. cit. note 25; National Oceanic and Atmospheric Administration (NOAA), *Natural Disaster Survey Report: July 1995 Heat Wave* (Silver Spring, MD: NOAA, December 1995); "Many of Last Year's Heat Deaths Were Preventable, Study Says," *Washington Post,* April 12, 1996.

37. A.J. Michael, *Planetary Overload: Global Environmental Change and the Health of the Human Species* (Cambridge: Cambridge University Press, 1993); C.E. Ewan et al., eds., *Health in the Greenhouse: The Medical and Environmental Health Effects of Global Climate Change* (Canberra: Australian Government Publishing Service, 1993); Houghton et al., eds., op. cit. note 1; Willem J.M. Martens et al., "Potential Impact of Global Climate Change on Malaria Risk," *Environmental Health Perspective,* May 1995; Anne Platt, *Infecting Ourselves: How Environmental and Social Disruptions Trigger Disease* (Washington, D.C.: Worldwatch Institute, 1996); William A. Sprigg, "Doctors Watch the Forecasts," *Nature*, February 15, 1996.

38. James P. Bruce, "Challenges of the Decade: Natural Disasters and Global Change," speech delivered at the Symposium on the World at Risk: Natural Hazards and Climate Change, Massachusetts Institute of Technology, Cambridge, MA, January 14-16, 1992; G.A. Berz, "Greenhouse Effects on Natural Catastrophes and Insurance," *The Geneva Papers on Risk Insurance, July 17, 1992*; NOAA quote from Karl et al., op. cit. note 17; Paul Simons, "Why Global Warming Could Take Britain by Storm," *New Scientist,* November 7, 1992; Craig R. Whitney, "Rhine Floods Worst in Century; 50,000 Homeless," *New York Times*, December 25, 1994.

39. Emanual estimate included in Doug Cogan, "Bracing for Bigger Storms," *Investor's Environmental Report*, Vol. 3, No. 1, 1993; see also Munich Re, *Windstorm*, Munich Re special publication (Munich, 1990); Graedel and Crutzen, op. cit. note 6.

40. Friedman calculation included in Cogan, op. cit. note 39; see also Munich Re, op. cit. note 39.

41. Robert C. Sheets, "Catastrophic Hurricanes May Become Frequent Events Along the United States East and Gulf Coasts," testimony before the U.S. Senate Government Affairs Committee, Washington, D.C., April 29, 1993; Greenpeace International, *The Climate Time Bomb: Signs of Climate Change from the Greenpeace Database* (Amsterdam, 1994), supplemented by "Update," March 1995; storm damage figures, including those in Table 1, are from Munich Re/Munchener Ruckversicherungs-Gesellschaft, "Fifty

Significant Loss Events of 1995," unpublished printout (Munich, May 10, 1996).

42. Greg Steinmetz, "Andrew's Toll: As Insurance Costs Soar, Higher Rates Loom," *Wall Street Journal,* January 6, 1993; E.N. Rappaport and R.B. Sheets, "A Meteorological Analysis of Hurricane Andrew," Lawrence S. Tait, ed., *Lessons of Hurricane Andrew,* excerpts from the 15th National Hurricane Conference, held inOrlando, FL, April 13-16, 1993 (Washington, D.C.: Federal Emergency Management Agency, 1993).

43. Dork L. Sahagian, Frank W. Schwartz, and David K. Jacobs, "Direct Anthropogenic Contributions to Sea Level Rise in the Twentieth Century," *Nature,* January 6, 1994; Malcolm W. Browne, "Most Precise Gauge Yet Points to Global Warming," *New York Times,* December 12, 1994; Houghton et al., eds., op. cit. note 1.

44. Watson et al., eds., op. cit. note 25.

45. Ibid.; Asian Development Bank, *Climate Change in Asia: The Thematic Overview* (Manila, 1994); Manuela Saragosa, "Floods Threaten Jakarta's Big Expansion Plan," *Financial Times,* January 13-14, 1996; impact on Egypt from Walter H. Corson, ed., *The Global Ecology Handbook: What You Can Do About the Environmental Crisis,* (Washington, D.C.: Global Tomorrow Coalition, 1990) as cited in Patricia Glick, *Global Warming: The High Cost of Inaction* (Washington, D.C.: Sierra Club, 1996); Rene Bowser et al., *Southern Exposure: Global Climate Change and Developing Countries* (Washington, D.C.: Center for Global Change, 1992); David E. Pitt, "Computer Vision of Global Warming: Hardest on Have-Nots," *New York Times,* January 18, 1994; Norman Myers and Jennifer Kent, *Environmental Exodus: An Emergent Crisis in the Global Arena* (Washington, D.C.: Climate Institute, 1995) as cited in Glick, op. cit. this note.

46. Colin D. Woodroffe, "Preliminary Assessment of the Vulnerability of Kiribati to Accelerated Sea Level Rise," in Joan O'Callahan, ed., *Global Climate Change and the Rising Challenge of the Sea,* Proceedings of the IPCC Workshop Held at Margarita Island, Venezuela, March 9-13, 1992 (Silver Spring, MD: NOAA, 1994).

47. Fankhauser, op. cit. note 33; Jeremy Leggett, ed., *Climate Change and the Financial Sector* (Munich: Gerling Akademie Verlag, 1996).

48. Figure 4 from Munich Re, op cit. note 41. Berlin insurance meeting description from author's observations, Berlin, March 26, 1995; H.R. Kaufmann, "Storm Damage Insurance—Quo Vadis?" unpublished paper produced by Swiss Re, Zurich, Switzerland, November 1990.

49. Gerhard A. Berz, "Global Warming and the Insurance Industry," *Interdisciplinary Science Reviews,* Vol. 18, No. 2, 1993.

50. William R. Cline, *The Economics of Global Warming* (Washington, D.C.: Institute of International Economics, 1992); Fankhauser, op. cit. note 33; Munich Re/Munchener Ruckversicherungs-Gesellschaft, op. cit. note 41.

51. North Korea's economic losses from Munich Re/Munchener Ruckversicherungs-Gesellschaft, op. cit. note 41.

52. *United Nations Framework Convention on Climate Change, Text* (Geneva: U.N. Environment Programme/World Meteorological Organization Information Unit on Climate Change, 1992); emissions from G. Marland, R.J. Andres, and T.A. Boden, "Global, Regional, and National CO2 Emission Estimates From Fossil Fuel Burning, Cement Production, and Gas Flaring: 1950-1992" (electronic database) (Oak Ridge, TN.: Carbon Dioxide Information Analysis Center, Oak Ridge National Laboratory, 1995), and from Worldwatch estimates based on ibid. and on British Petroleum (BP), *BP Statistical Review of World Energy* (London: Group Media & Publications, 1995).

53. Emission figures in this paper are expressed in metric tons of carbon, even when the term carbon dioxide is used. As fossil fuels are burned, the carbon in them is combined with oxygen to become carbon dioxide which has a greater molecular weight. Carbon emission figures, including those in Figure 5, are from Marland et al., op. cit. note 52, and Worldwatch estimates based on ibid. and on BP, op. cit. note 52; Houghton et al., eds., op. cit. note 1.

54. Worldwatch estimates based on Organisation for Economic Co-operation and Development (OECD) and International Energy Agency, *Energy Balances of OECD Countries 1960-79*; ibid., *World Energy Statistics and Balances 1971-1987*; and ibid., *Energy Statistics and Balances of Non-OECD Countries 1991-1992* (Paris: OECD, 1991, 1989, and 1994 respectively).

55. Per capita carbon emission figures, including those in Table 2, are based on Marland et al., op. cit. note 52; Worldwatch estimates based on ibid., on BP, op. cit. note 52, and on Population Reference Bureau, *1994 World Population Data Sheet* (Washington, D.C., 1994).

56. Ibid.; GNP data for 1993 adjusted for purchasing power parity from *The World Bank Atlas 1995* (Washington, D.C.: World Bank, 1995).

57. *United Nations Framework Convention on Climate Change,* op. cit. note 52.

58. President William J. Clinton and Vice President Albert Gore, Jr., *The Climate Change Action Plan* (Washington, D.C.: The White House, 1993); William K. Stevens, "U.S. Prepares to Unveil Blueprint for Reducing Heat-Trapping Gases," *New York Times*, October 12, 1993; Gary Lee, "Sorting Out the Sources of Greenhouse Gases," *Washington Post,* October 26, 1993; Joel Darmstadter, "The U.S. Climate Change Action Plan: Challenges and Prospects," *Resources*, Winter 1995.

59. Clinton and Gore, op. cit. note 58; Daniel Lashof, senior scientist at Natural Resources Defense Council (NRDC), as quoted in National Public Radio's "All Things Considered," December 11, 1995.

60. Energy Information Administration (EIA), *Annual Energy Outlook 1996* (Washington, D.C., 1996); U.S. Climate Action Network (US CAN) and Climate Network Europe (CNE), *Independent NGO Evaluations of National Plans for Climate Change Mitigation: OECD Countries, Third Review, January 1995* (Washington, D.C.: U.S. Climate Action Network, 1995).

61. Commission of the European Communities *Second Evaluation of National Programmes Under the Monitoring Mechanism of Community CO2 and Other Greenhouse Gas Emissions* (Brussels, March 14, 1996); "Commission Report Finds Current Policies on CO2 Emissions 'Clearly Insufficient'," *International Environment Reporter,* April 3, 1996.

62. Chancellor Helmut Kohl, speech delivered at the First Conference of the Parties to the United Nations Framework Convention on Climate Change, Berlin, April 5, 1995; US CAN and CNE, op. cit. note 60; International Energy Agency (IEA), *Climate Change Policy Initiatives, Volume 1: OECD Countries* (Paris: OECD, 1994); emissions from Marland et al., op. cit. note 52, and Worldwatch estimates based on ibid. and on BP, op. cit. note 52.

63. US CAN and CNE, op. cit. note 60; IEA, op. cit. note 62; Christopher Flavin, "Wind Power Soars," in Lester R. Brown, Nicholas Lenssen, and Hal Kane, *Vital Signs 1995* (New York: W.W. Norton, 1995).

64. Nathaniel Nash, "German High Court Bans Energy Subsidy on Utility Bills," *New York Times,* December 8, 1994; reverse carbon tax estimate by Worldwatch based on ibid. and on BP, op. cit. note 52; "Accord on German Coal Subsidies Highlights Bonn Coalition Rift," *European Energy Report,* March 17, 1995; Wilfrid Bach, "Coal Policy and Climate Protection: Can the Tough German CO2 Reduction Target Be Met By 2005?" *Energy Policy,* Vol. 23, No. 1, 1995; Stephan Singer and Manfred Treber, *Report to the Commission on Sustainable Development,* German NGO Forum on Environment and Development, Task Force on Climate, (Bonn, March 1996).

65. Singer and Treber, op. cit. note 64; German Bundestas Enquete Commission on Protecting the Earth's Atmosphere *Mobility and Climate: Developing Environmentally Sound Transport Policy Concepts* (Bonn, Economica Verlag, 1995).

66. Gasoline tax from International Energy Agency (IEA), *Energy Prices and Taxes, First Quarter, 1995* (Paris: OECD, 1995); carbon equivalent estimate by Worldwatch; "Industry Announces Voluntary Strategy for Cut CO2 Emissions 20 Percent by 2005," *International Environment Reporter,* April 3, 1996.

67. Singer and Treber, op. cit. note 64.

68. Danish Ministry of the Environment, *Climate Protection in Denmark* (Copenhagen: Danish Environmental Protection Agency, 1994); U.S. Climate Action Network and Climate Network Europe, op. cit. note 60; International Energy Agency, op. cit. note 62.

69. US CAN and CNE, op. cit. note 60; IEA, op. cit. note 62; "Memorandum on Energy Use Targets 33 Percent Improvement in Efficiency by 2020," *International Environment Reporter*, January 24, 1996; CNE, *CNE Report on EU Country-By-Country Situations* (Brussels, March 1996); Commission of the European Communities *Second Evaluation of National Programmes Under the Monitoring Mechanism of Community CO2 and Other Greenhouse Gas Emissions* (Brussels, March 14, 1996).

70. Commission of the European Communities, op. cit. note 69.

71. US CAN and CNE, op. cit. note 60; IEA, op. cit. note 62; Dwight Van Winkle, "Japan's CO2 Emissions Rise Post-2000," *Climate Forum* (electronic Econet conference), June 12, 1995; "Agency Planning More Comprehensive Laws on Global Warming," *Foreign Broadcast Information Service (FBIS) Report,* December 27, 1995.

72. Marland et al., op. cit. 52; Worldwatch estimates based on ibid., and on BP, op. cit. note 52; US CAN and CNE, *Independent NGO Evaluations of National Plans for Climate Change Mitigation: Central and Eastern Europe, First Review, January 1995* (Washington, D.C.: US CAN, 1995).

73. Marland et al., op. cit. 52; Worldwatch estimates based on ibid., and on BP, op. cit. note 52.

74. Ibid.; Jessica Hamburger, *China's Energy and Environment in the Roaring Nineties: A Policy Primer* (Washington, D.C.: Pacific Northwest Laboratory, 1995).

75. *United Nations Framework Convention on Climate Change,* op. cit. note 52.

76. "Little Progress at Climate Convention Meeting," *Environment Watch: Western Europe,* March 15, 1996.

77. International Energy Agency (IEA), *World Energy Outlook: 1996 Edition* (Paris: OECD, 1996).

78. *United Nations Framework Convention on Climate Change,* op. cit. note 52.

79. First Conference of the Parties to the Framework Convention on Climate Change, "Conclusion of Outstanding Issues and Adoption of Decisions: Proposal on Agenda Item 5 (a) (iii) Submitted by the President of the Conference; Review of the Adequacy of Article 4, Paragraph 2 (a) and

(b) of the Convention, Including Proposals Related to a Protocol and Decisions on Follow-up," Berlin, April 7, 1995; Michael Zammit Cutajar, "Berlin Mandate Opens Way for Post-2000 Commitments," *United Nations Climate Change Bulletin*, Issue 7, 2nd Quarter, 1995.

80. F.R. Rijsberman and R.J. Stewart, eds., "Targets and Indicators of Climate Change," Stockholm Environment Institute, 1990; German Bundestag Enquete Commission on Protecting the Earth's Atmosphere, *Climate Change—A Threat to Global Development* (Bonn, Economica Verlag, 1992); German Advisory Council on Global Change, *Scenario for the derivation of Global CO2 Reduction Targets and Implementation Strategies* (Bremerhaven, 1995).

81. Houghton et al., eds., op. cit. note 1; Wilfrid Bach and Atul K. Jain, "Toward Climate Conventions: Scenario Analysis for a Climatic Protection Policy," *Ambio*, November 1991; Joseph Alcamo and Eric Kreileman, *The Global Climate System: Near Term Action for Long Term Protection* (background paper) (the Netherlands: National Institute of Publice Health and the Environment (RIVM), February, 1996); Figure 6 data are from T.M.L. Wigley, R. Richels, and J.A. Edmonds, "Economic and Environmental Choices in the Stabilization of Atmospheric CO2 Concentrations," *Nature*, January 18, 1996.

82. Alcamo and Kreileman, op. cit. note 81; Figure 7 data are from Wigley et al., op. cit. note 81; the carbon emission figures in Figure 7 include carbon emissions from deforestation.

83. Alcamo and Kreileman, op. cit. note 82.

84. Hal Kane, "Sulfur and Nitrogen Emissions Flat," in Lester Brown, Christopher Flavin, and Hal Kane, *Vital Signs 1996* (New York: W.W. Norton, 1996); Janusz Cofala, "Modeling Acid Rain in Southeast Asia," *IIASA Options*, Winter 1993.

85. United Nations, *Long-range World Population Projections: Two Centuries of World Population Growth, 1950-2150* (New York, 1992).

86. Christopher Flavin and Nicholas Lenssen, *Power Surge: Guide to the Coming Energy Revolution* (New York: W.W. Norton, 1994); Thomas B. Johansson, et al., "Renewable Fuels and Electricity for a Growing World Economy: Defining and Achieving the Potential," in Laurie Burnham, ed., *Renewable Energy: Sources for Fuels and Electricity* (Washington, D.C.: Island Press, 1993); Florentin Krause, *Energy Policy in the Greenhouse* (El Cerrito, CA.: International Project for Sustainable Energy Paths, 1993).

87. Jesse H. Ausubel, "Technical Progress and Climatic Change," *Energy Policy*, April/May 1995; World Energy Council (WEC) and International Institute for Applied Systems Analysis (IIASA), *Global Energy Perspectives to 2050 and Beyond* (London: World Energy Council, 1995).

88. J.M. Martin, "L'intensité énergétique de l'activité économique dans les pays industrialisés: Les évolutions de très longue période livrent-elles des enseignements utiles?" *Économies et Sociétés*, No. 4, 1988; Energy Information Administration (EIA), *Monthly Energy Review: May 1996* (Washington, D.C., 1996).

89. Lee Schipper and Stephen Meyers, *Energy Efficiency and Human Activity: Past Trends, Future Prospects* (New York: Cambridge University Press, 1992); Hamburger, op. cit. note 74; Alcamo and Kreileman, op. cit. note 81.

90. Charles Duell quoted in Christopher Cerf and Victor Navasky, *The Experts Speak: The Definitive Compendium of Authoritative Disinformation* (New York, NY: Pantheon Books, 1984).

91. Flavin and Lenssen, op. cit. note 86; WEC and IIASA, op. cit. note 87; Robert H. Williams, "The LESS Constructions of the IPCC," Center for Energy and Environmental Studies, Princeton University, September 1995.

92. IEA, op. cit. note 77; WEC and IIASA, op. cit. note 87.

93. Christopher Flavin, "Power Shock," *World Watch*, January/February, 1996.

94. Brown, Flavin, and Kane, op. cit. note 84.

95. Flavin and Lenssen, op. cit. note 86; WEC and IIASA, op. cit. note 87; Denis Hayes, *Rays of Hope* (New York: W.W. Norton, 1977).

96. David G. Howell, ed., *The Future of Energy Gases* (Washington, D.C.: U.S. Geological Survey, 1993); Worldwatch Institute calculation, based on the fact that U.S coal plants emit 32 percent of U.S. emissions according to EIA, op. cit. note 60; gas-fired cogeneration could reduce these emissions by 80 percent.

97. EIA, op. cit. note 60.

98. Peter Kassler, *Energy for Development* (London: Royal Shell International Petroleum Company, 1994).

99. WEC and IIASA, op. cit. note 87.

100. Ibid.

101. William R. Cline, *The Economics of Global Warming* (Washington, D.C.: Institute for International Economics, 1992); J. Sathaye and J. Christensen, "Methods for the Economic Evaluation of Greenhouse Gas Mitigation Options," *Energy Policy*, Special Issue, November 1994; R. Richels and J. Edmonds, "The Economics of Stabilizing Atmospheric CO2 Concentrations," *Energy Policy*, April/May 1995; Sylvia M. Rothen, "The Greenhouse Effect in Economic Modeling," unpublished paper produced for the Human Ecology Group, Swiss Federal Institute for Environmental Science and Technology, Dubendorf, Switzerland, October 1994.

102. A.J. McMichael, "Carbon Dioxide Emissions," *Nature*, February 29, 1996; Cline, op. cit., note 101; J. Bruce, Hoesung Lee, and E. Haites, eds., *Climate Change 1995: Economic and Social Dimensions of Climate Change, Contribution of Working Group III to the Second Assessment Report of the Intergovernmental Panel on Climate Change* (New York: Cambridge University Press, 1996).

103. John Palmisano, "The U.S. Acid Deposition Control Program," unpublished draft, April 10, 1996; Greenpeace, "A Technology that Conquers the World! Hydrocarbons-Hightech Refrigeration" (a 14-page paper) "Greenfreeze: The World's First CFC- and HFC-free Household Refrigerators, and a Worldwide Success for Natural Gases," Hamburg, Germany, May 1994; Ozone Action, "Hydrocarbons: An Alternative to Ozone Depleting Chemicals in Refrigerators" (information sheet) Washington, D.C., November, 1995.

104. "Europe's Most Modern Combined-Cycle Plant," *Electricity International*, June/July 1993.

105. Flavin and Lenssen, op. cit. note 86.

106. Wigley et al., op. cit. 81; Fred Pearce, "Sit Tight for 30 Years, Argues Climate Guru," *New Scientist*, January 20, 1996.

107. Michael Grubb, "Technologies, Energy Systems, and the Timing of CO2 Emissions Abatement: An Overview of Economic Issues," presentation to a workshop at the Center for Global Change, University of Maryland, February 8-10, 1996.

108. First Conference of the Parties to the Framework Convention on Climate Change, op. cit. note 79.

109. Stephen Kinzer, "U.N. Parley Delegates Back Talks on Global Warming," *New York Times*, April 8, 1995; Timothy Noah, "Rio Summit Group Sets Date for Limits on Some Emissions," *Wall Street Journal*, April 10, 1995; Michael Grubb, "Viewpoint: The Berlin Climate Conference; Shifting Alliances Break Political Deadlock," *EC Energy Monthly*, April 21, 1995.

110. Karan Capoor and Annie Petsonk, "The Climate Summit: From Rio to Berlin and Beyond," *Hotline*, June 1995; Grubb, op. cit. note 109; Gummer quoted in Seth Dunn, "The Berlin Climate Change Summit: Implications for International Environmental Law," *International Environment Reporter*, May 31, 1995.

111. "Report of the Second Meeting of the Subsidiary Bodies of the UN

Framework Convention on Climate Change (27 February-4 March, 1996)," *Earth Negotiations Bulletin*, March 11, 1996.

112. Christopher Flavin and Odil Tunali, "Getting Warmer: Looking for a Way Out of the Climate Impasse," *World Watch*, March/April, 1995.

113. David G. Victor and Julian E. Salt, "Keeping the Climate Treaty Relevant," *Nature*, January 26, 1995; David G. Victor, "The Uses and Design of Targets in International Environmental Agreements," paper presented at a conference on global change, Aspen Global Change Institute, Aspen, Colo., August 14, 1995.

114. Bruce N. Stram, "A Carbon Tax Strategy for Global Climate Change," in Henry Lee, ed., *Shaping National Responses to Climate Change: A Post-Rio Guide* (Washington, D.C.: Island Press, 1995); David Malin Roodman, "Harnessing the Market for the Environment," in *State of the World 1996* (New York: W.W. Norton, 1996).

115. Frank Muller, "Mitigating Climate Change: The Case for Energy Taxes," *Environment*, Vol. 38, No. 2, 1996.

116. Palmisano, op. cit. note 103; Dan Seligman, *Air Pollution Emissions Trading: Opportunity or Scam?* (Washington, D.C.: Sierra Club, 1995).

117. Robert W. Hahn and Robert N. Stavins, "Trading in Greenhouse Permits: A Critical Examination of Design and Implementation Issues," in Lee, ed., op. cit. note 114; Randall Spalding-Fecher, "Joint Implementation and Energy Efficiency," unpublished course paper written at the Fletcher School of Law and Diplomacy, May 6, 1996.

118. Tim Jackson, "Joint Implementation and Cost-Effectiveness Under the Framework Convention on Climate Change," *Energy Policy*, February 1995; "Conflicts of Interest on the Greenhouse," *Nature*, April 6, 1995; Sierra Club, "Risky Business: Why Joint Implementation is the Wrong Approach to Global Warming Policy," Washington, D.C., April 1995; Jyoti K. Parikh, "Joint Implementation and North-South Cooperation for Climate Change," *International Environmental Affairs*, Winter 1995; "Eight New USIJI Projects Announced at White House Ceremony," International Partnerships Report, Winter 1996.

119. "Eight New USIJI Projects Announced at White House Ceremony," *International Partnerships Report*, Winter 1996.

120. EIA, op. cit. note 60; estimates of tree planting required are by Worldwatch Institute based on Robert N. Stavins, *The Costs of Carbon Sequestration: A Revealed Preference Approach* (Cambridge, MA.: John F. Kennedy School of Government, 1995).

121. Kilaparti Ramakrishna, *Criteria for Joint Implementation Under the*

Framework Convention on Climate Change (Woods Hole, MA.: Woods Hole Research Center, 1994).

122. John Palmisano, "A Proposal to Evaluate Joint Implementation Before the 1997 Tokyo Conference of Parties," unpublished paper prepared for the Business Councils for a Sustainable Energy Future, May 1996.

123. Fiona Mullins, "Demand Side Efficiency: Product Standards" (draft) (Paris: OECD, March 25, 1996).

124. Ronald van de Krol, "Partners in Grime," *Financial Times*, May 1, 1996; Larry Mansueti, "The Electric Utility Industry/DOE Climate Challenge Program," U.S. Department of Energy, May 1996; Lee Solsberg and Peter Wiederkehr, "Voluntary Approaches for Energy-Related CO2 Abatement," *OECD Observer*, October/November, 1995; Wayne Tusa, "Understanding and Implementing ISO 14000, *Developments in Federal and State Law,* April 1996.

125. U.S. Environmental Protection Agency, *Green Light: Fourth Annual Report* (Washington, D.C., 1995).

126. Global Environment Facility (GEF), *Quarterly Operational Report* (Washington, D.C., December 1995).

127. Less than 5 percent estimate from Lara Helfer, International Institute for Energy Conservation, private communication, May 14, 1996. The World Bank, *The World Bank and the UN Framework Convention on Climate Change* (Washington, D.C., 1995); The World Bank, *Mainstreaming the Environment* (Washington, D.C., 1995).

128. Gelbspan, op. cit. note 14; Michael Grubb, "The Berlin Climate Conference: Outcome and Implications," *FEEM Newsletter,* No. 2, 1995; Grubb, op. cit. note 109; Dunn, op. cit. note 110.

129. Leggett, ed., op. cit. note 47.

130. Lloyds representative quoted in Jeremy Leggett, "A Looming Capital Crisis for Oil? Taking Bearings in the Greenhouse in a Post Brent-Spar World," presentation before the Aspen Environmental Roundtable, Aspen, CO, September 18, 1995.

131. Dunn, op. cit. note 110; Carol Werner and Jennifer Morgan, "Cities Endorse AOSIS Protocol," *ECO,* March 30, 1995; International Council for Local Environmental Initiatives (ICLEI), *Saving the Climate—Saving the Cities* (Toronto: ICLEI, 1993).

132. Grubb, op. cit. notes 109 and 128; Dunn, op. cit. note 110.

133. Ibid.; Asian Development Bank, op. cit. note 45.

134. Grubb, op. cit. notes 109 and 128.

135. Hilary F. French, "Implementing the Ozone Treaty in Developing Countries: Is the Global Partnership Working?" unpublished course paper written at the Fletcher School of Law and Diplomacy, unpublished, May 7, 1995; Richard Elliot Benedick, *Ozone Diplomacy: New Directions in Safeguarding the Planet* (Cambridge, MA: Harvard University Press, 1991).

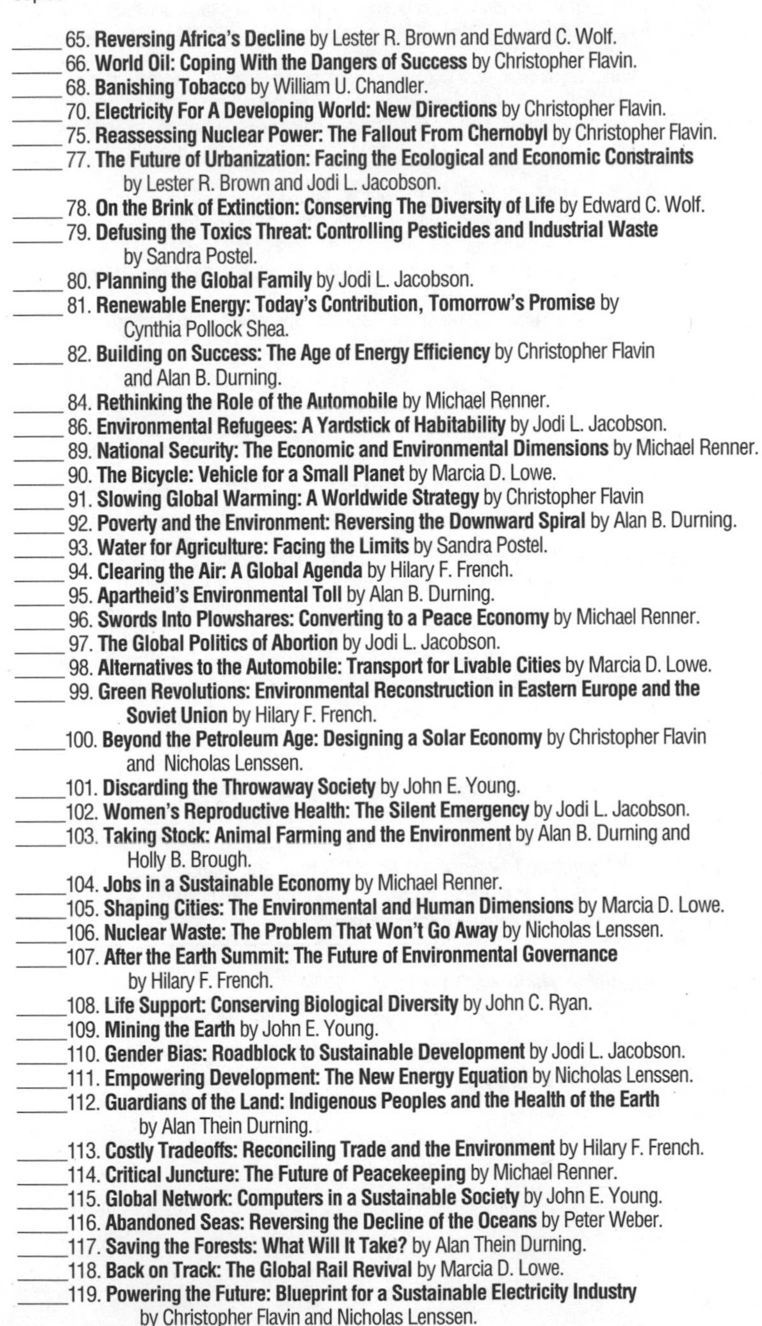

PUBLICATION ORDER FORM

No. of
Copies

_____ 65. **Reversing Africa's Decline** by Lester R. Brown and Edward C. Wolf.
_____ 66. **World Oil: Coping With the Dangers of Success** by Christopher Flavin.
_____ 68. **Banishing Tobacco** by William U. Chandler.
_____ 70. **Electricity For A Developing World: New Directions** by Christopher Flavin.
_____ 75. **Reassessing Nuclear Power: The Fallout From Chernobyl** by Christopher Flavin.
_____ 77. **The Future of Urbanization: Facing the Ecological and Economic Constraints** by Lester R. Brown and Jodi L. Jacobson.
_____ 78. **On the Brink of Extinction: Conserving The Diversity of Life** by Edward C. Wolf.
_____ 79. **Defusing the Toxics Threat: Controlling Pesticides and Industrial Waste** by Sandra Postel.
_____ 80. **Planning the Global Family** by Jodi L. Jacobson.
_____ 81. **Renewable Energy: Today's Contribution, Tomorrow's Promise** by Cynthia Pollock Shea.
_____ 82. **Building on Success: The Age of Energy Efficiency** by Christopher Flavin and Alan B. Durning.
_____ 84. **Rethinking the Role of the Automobile** by Michael Renner.
_____ 86. **Environmental Refugees: A Yardstick of Habitability** by Jodi L. Jacobson.
_____ 89. **National Security: The Economic and Environmental Dimensions** by Michael Renner.
_____ 90. **The Bicycle: Vehicle for a Small Planet** by Marcia D. Lowe.
_____ 91. **Slowing Global Warming: A Worldwide Strategy** by Christopher Flavin
_____ 92. **Poverty and the Environment: Reversing the Downward Spiral** by Alan B. Durning.
_____ 93. **Water for Agriculture: Facing the Limits** by Sandra Postel.
_____ 94. **Clearing the Air: A Global Agenda** by Hilary F. French.
_____ 95. **Apartheid's Environmental Toll** by Alan B. Durning.
_____ 96. **Swords Into Plowshares: Converting to a Peace Economy** by Michael Renner.
_____ 97. **The Global Politics of Abortion** by Jodi L. Jacobson.
_____ 98. **Alternatives to the Automobile: Transport for Livable Cities** by Marcia D. Lowe.
_____ 99. **Green Revolutions: Environmental Reconstruction in Eastern Europe and the Soviet Union** by Hilary F. French.
_____100. **Beyond the Petroleum Age: Designing a Solar Economy** by Christopher Flavin and Nicholas Lenssen.
_____101. **Discarding the Throwaway Society** by John E. Young.
_____102. **Women's Reproductive Health: The Silent Emergency** by Jodi L. Jacobson.
_____103. **Taking Stock: Animal Farming and the Environment** by Alan B. Durning and Holly B. Brough.
_____104. **Jobs in a Sustainable Economy** by Michael Renner.
_____105. **Shaping Cities: The Environmental and Human Dimensions** by Marcia D. Lowe.
_____106. **Nuclear Waste: The Problem That Won't Go Away** by Nicholas Lenssen.
_____107. **After the Earth Summit: The Future of Environmental Governance** by Hilary F. French.
_____108. **Life Support: Conserving Biological Diversity** by John C. Ryan.
_____109. **Mining the Earth** by John E. Young.
_____110. **Gender Bias: Roadblock to Sustainable Development** by Jodi L. Jacobson.
_____111. **Empowering Development: The New Energy Equation** by Nicholas Lenssen.
_____112. **Guardians of the Land: Indigenous Peoples and the Health of the Earth** by Alan Thein Durning.
_____113. **Costly Tradeoffs: Reconciling Trade and the Environment** by Hilary F. French.
_____114. **Critical Juncture: The Future of Peacekeeping** by Michael Renner.
_____115. **Global Network: Computers in a Sustainable Society** by John E. Young.
_____116. **Abandoned Seas: Reversing the Decline of the Oceans** by Peter Weber.
_____117. **Saving the Forests: What Will It Take?** by Alan Thein Durning.
_____118. **Back on Track: The Global Rail Revival** by Marcia D. Lowe.
_____119. **Powering the Future: Blueprint for a Sustainable Electricity Industry** by Christopher Flavin and Nicholas Lenssen.

_____120. **Net Loss: Fish, Jobs, and the Marine Environment** by Peter Weber.
_____121. **The Next Efficiency Revolution: Creating a Sustainable Materials Economy** by John E. Young and Aaron Sachs.
_____122. **Budgeting for Disarmament: The Costs of War and Peace** by Michael Renner.
_____123. **High Priorities: Conserving Mountain Ecosystems and Cultures** by Derek Denniston.
_____124. **A Building Revolution: How Ecology and Health Concerns Are Transforming Construction** by David Malin Roodman and Nicholas Lenssen.
_____125. **The Hour of Departure: Forces That Create Refugees and Migrants** by Hal Kane.
_____126. **Partnership for the Planet: An Environmental Agenda for the United Nations** by Hilary F. French.
_____127. **Eco-Justice: Linking Human Rights and the Environment** by Aaron Sachs.
_____128. **Imperiled Waters, Impoverished Future: The Decline of Freshwater Ecosystems** by Janet N. Abramovitz.
_____129. **Infecting Ourselves: How Environmental and Social Disruptions Trigger Disease** by Anne E. Platt.
_____130. **Climate of Hope: New Strategies for Stabilizing the World's Atmosphere** by Christopher Flavin and Odil Tunali.

_____ **Total Copies**

Single Copy: $5.00 • 2–5: $4.00 ea. • 6–20: $3.00 ea. • 21 or more: $2.00 ea.
Call Director of Communication at (202) 452-1999 to inquire about discounts on larger orders.

☐ **Membership in the Worldwatch Library: $30.00 (international airmail $45.00)**
The paperback edition of our 250-page "annual physical of the planet," *State of the World,* plus all Worldwatch Papers released during the calendar year.

☐ **Subscription to *World Watch* magazine: $20.00 (international airmail $35.00)**
Stay abreast of global environmental trends and issues with our award-winning, eminently readable bimonthly magazine.

☐ **Worldwatch Database Disk Subscription: One year for $89**
Includes current global agricultural, energy, economic, environmental, social, and military indicators from all current Worldwatch publications. Includes a mid-year update, and *Vital Signs* and *State of the World* as they are published. Can be used with Lotus 1-2-3, Quattro Pro, Excel, SuperCalc and many other spreadsheets.
Check one: _____high-density IBM-compatible or _____Macintosh

Make check payable to Worldwatch Institute
1776 Massachusetts Avenue, N.W., Washington, D.C. 20036-1904 USA

Please include $3 postage and handling for non-subscription orders.

Enclosed is my check for U.S. $________
AMEX ☐ VISA ☐ Mastercard ☐ ________________________________
Card Number Expiration Date

____
name **daytime phone #**

____
address

____
city **state** **zip/country**

Phone: (202) 452-1999 Fax: (202) 296-7365 E-Mail: wwpub@worldwatch.org

WWP

PUBLICATION ORDER FORM

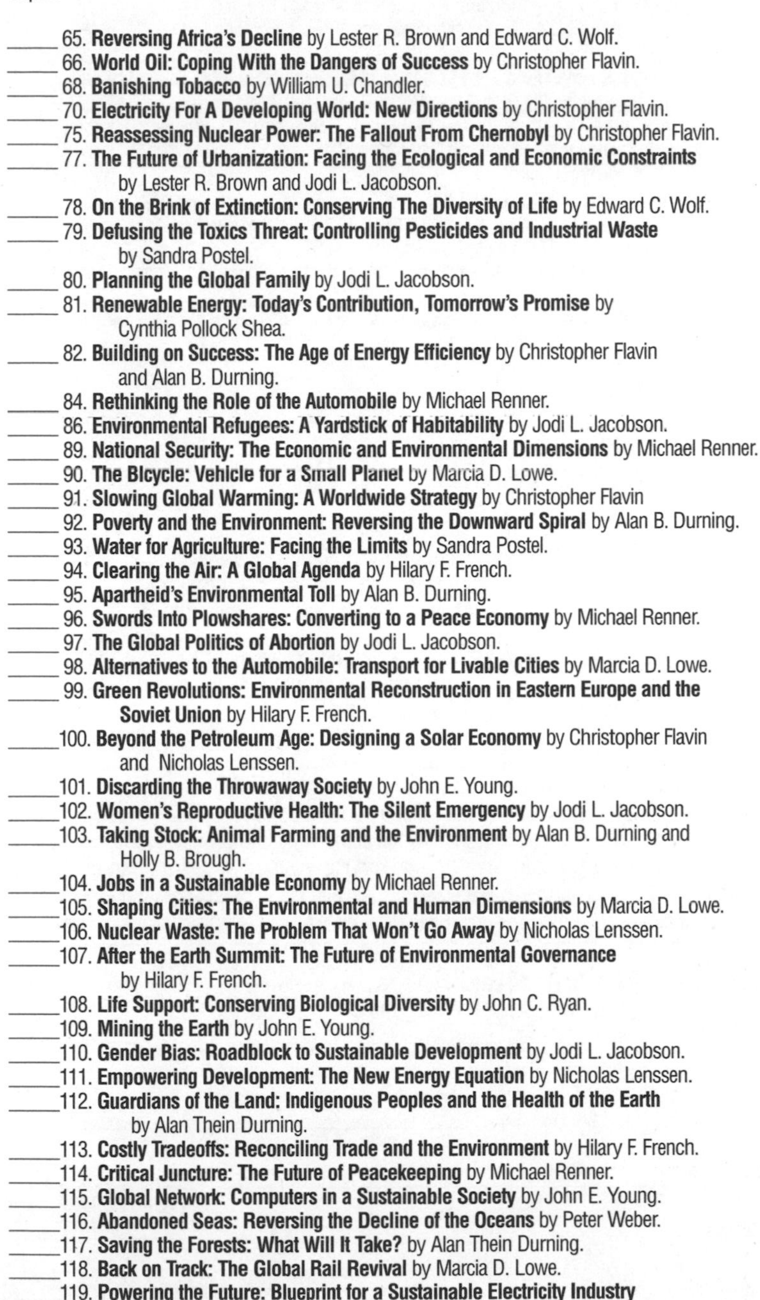

No. of
Copies

_____ 65. **Reversing Africa's Decline** by Lester R. Brown and Edward C. Wolf.
_____ 66. **World Oil: Coping With the Dangers of Success** by Christopher Flavin.
_____ 68. **Banishing Tobacco** by William U. Chandler.
_____ 70. **Electricity For A Developing World: New Directions** by Christopher Flavin.
_____ 75. **Reassessing Nuclear Power: The Fallout From Chernobyl** by Christopher Flavin.
_____ 77. **The Future of Urbanization: Facing the Ecological and Economic Constraints** by Lester R. Brown and Jodi L. Jacobson.
_____ 78. **On the Brink of Extinction: Conserving The Diversity of Life** by Edward C. Wolf.
_____ 79. **Defusing the Toxics Threat: Controlling Pesticides and Industrial Waste** by Sandra Postel.
_____ 80. **Planning the Global Family** by Jodi L. Jacobson.
_____ 81. **Renewable Energy: Today's Contribution, Tomorrow's Promise** by Cynthia Pollock Shea.
_____ 82. **Building on Success: The Age of Energy Efficiency** by Christopher Flavin and Alan B. Durning.
_____ 84. **Rethinking the Role of the Automobile** by Michael Renner.
_____ 86. **Environmental Refugees: A Yardstick of Habitability** by Jodi L. Jacobson.
_____ 89. **National Security: The Economic and Environmental Dimensions** by Michael Renner.
_____ 90. **The Bicycle: Vehicle for a Small Planet** by Marcia D. Lowe.
_____ 91. **Slowing Global Warming: A Worldwide Strategy** by Christopher Flavin
_____ 92. **Poverty and the Environment: Reversing the Downward Spiral** by Alan B. Durning.
_____ 93. **Water for Agriculture: Facing the Limits** by Sandra Postel.
_____ 94. **Clearing the Air: A Global Agenda** by Hilary F. French.
_____ 95. **Apartheid's Environmental Toll** by Alan B. Durning.
_____ 96. **Swords Into Plowshares: Converting to a Peace Economy** by Michael Renner.
_____ 97. **The Global Politics of Abortion** by Jodi L. Jacobson.
_____ 98. **Alternatives to the Automobile: Transport for Livable Cities** by Marcia D. Lowe.
_____ 99. **Green Revolutions: Environmental Reconstruction in Eastern Europe and the Soviet Union** by Hilary F. French.
_____100. **Beyond the Petroleum Age: Designing a Solar Economy** by Christopher Flavin and Nicholas Lenssen.
_____101. **Discarding the Throwaway Society** by John E. Young.
_____102. **Women's Reproductive Health: The Silent Emergency** by Jodi L. Jacobson.
_____103. **Taking Stock: Animal Farming and the Environment** by Alan B. Durning and Holly B. Brough.
_____104. **Jobs in a Sustainable Economy** by Michael Renner.
_____105. **Shaping Cities: The Environmental and Human Dimensions** by Marcia D. Lowe.
_____106. **Nuclear Waste: The Problem That Won't Go Away** by Nicholas Lenssen.
_____107. **After the Earth Summit: The Future of Environmental Governance** by Hilary F. French.
_____108. **Life Support: Conserving Biological Diversity** by John C. Ryan.
_____109. **Mining the Earth** by John E. Young.
_____110. **Gender Bias: Roadblock to Sustainable Development** by Jodi L. Jacobson.
_____111. **Empowering Development: The New Energy Equation** by Nicholas Lenssen.
_____112. **Guardians of the Land: Indigenous Peoples and the Health of the Earth** by Alan Thein Durning.
_____113. **Costly Tradeoffs: Reconciling Trade and the Environment** by Hilary F. French.
_____114. **Critical Juncture: The Future of Peacekeeping** by Michael Renner.
_____115. **Global Network: Computers in a Sustainable Society** by John E. Young.
_____116. **Abandoned Seas: Reversing the Decline of the Oceans** by Peter Weber.
_____117. **Saving the Forests: What Will It Take?** by Alan Thein Durning.
_____118. **Back on Track: The Global Rail Revival** by Marcia D. Lowe.
_____119. **Powering the Future: Blueprint for a Sustainable Electricity Industry** by Christopher Flavin and Nicholas Lenssen.

_____120. **Net Loss: Fish, Jobs, and the Marine Environment** by Peter Weber.
_____121. **The Next Efficiency Revolution: Creating a Sustainable Materials Economy** by John E. Young and Aaron Sachs.
_____122. **Budgeting for Disarmament: The Costs of War and Peace** by Michael Renner.
_____123. **High Priorities: Conserving Mountain Ecosystems and Cultures** by Derek Denniston.
_____124. **A Building Revolution: How Ecology and Health Concerns Are Transforming Construction** by David Malin Roodman and Nicholas Lenssen.
_____125. **The Hour of Departure: Forces That Create Refugees and Migrants** by Hal Kane.
_____126. **Partnership for the Planet: An Environmental Agenda for the United Nations** by Hilary F. French.
_____127. **Eco-Justice: Linking Human Rights and the Environment** by Aaron Sachs.
_____128. **Imperiled Waters, Impoverished Future: The Decline of Freshwater Ecosystems** by Janet N. Abramovitz.
_____129. **Infecting Ourselves: How Environmental and Social Disruptions Trigger Disease** by Anne E. Platt.
_____130. **Climate of Hope: New Strategies for Stabilizing the World's Atmosphere** by Christopher Flavin and Odil Tunali.

_____ **Total Copies**

Single Copy: $5.00 • 2–5: $4.00 ea. • 6–20: $3.00 ea. • 21 or more: $2.00 ea.
Call Director of Communication at (202) 452-1999 to inquire about discounts on larger orders.

☐ **Membership in the Worldwatch Library: $30.00 (international airmail $45.00)**
The paperback edition of our 250-page "annual physical of the planet," *State of the World,* plus all Worldwatch Papers released during the calendar year.

☐ **Subscription to *World Watch* magazine: $20.00 (international airmail $35.00)**
Stay abreast of global environmental trends and issues with our award-winning, eminently readable bimonthly magazine.

☐ **Worldwatch Database Disk Subscription: One year for $89**
Includes current global agricultural, energy, economic, environmental, social, and military indicators from all current Worldwatch publications. Includes a mid-year update, and *Vital Signs* and *State of the World* as they are published. Can be used with Lotus 1-2-3, Quattro Pro, Excel, SuperCalc and many other spreadsheets.
Check one: _____high-density IBM-compatible or _____Macintosh

Make check payable to Worldwatch Institute
1776 Massachusetts Avenue, N.W., Washington, D.C. 20036-1904 USA

Please include $3 postage and handling for non-subscription orders.

Enclosed is my check for U.S. $________

AMEX ☐ VISA ☐ Mastercard ☐ ______________________________
Card Number Expiration Date

____
name **daytime phone #**

____
address

____
city **state** **zip/country**

Phone: (202) 452-1999 Fax: (202) 296-7365 E-Mail: wwpub@worldwatch.org

WWP